Lionel ELLA

Local development and cooperation policies

Lionel ELLA

Local development and cooperation policies

in Cameroon's communes in the face of increasing decentralization

ScienciaScripts

Imprint

Cover image: www.ingimage.com

This book is a translation from the original published under ISBN 978-620-6-71341-8.

Publisher:
Sciencia Scripts
is a trademark of
Dodo Books Indian Ocean Ltd. and OmniScriptum S.R.L publishing group

120 High Road, East Finchley, London, N2 9ED, United Kingdom
Str. Armeneasca 28/1, office 1, Chisinau MD-2012, Republic of Moldova, Europe
Printed at: see last page
ISBN: 978-620-7-72754-4

Local development and cooperation policies in Cameroonian communes tested by the strengthening of decentralization

By

Lionel ELLA

2024

DEDICATION

A

My late Mom

ACKNOWLEDGEMENTS

Our sincere thanks go to :

- Dr Landry NGONO TSIMI for his guidance and invaluable advice;
- Professor Paul Elvic BATCHOM, for his availability;
- our parents Mr. and Mrs. ANDJONGO who have always shown their love and affection;
- our parents Mr. and Mrs. ESSISSIMA for leadership
- Mr. Yannick Martial AYISSI ELOUNDOU, Mayor of the Yaoundé II District Commune, for allowing us to carry out this research work in his prestigious community;
- Mrs RAISSA BETODEN, Head of the Cooperation and Local Development Department at the Yaoundé II Commune d'Arrondissement, for her professional and maternal welcome.
- Mr OTTOU, Head of the Legal Affairs Department at the Yaoundé II District Council, for his invaluable encouragement;
- Mrs Salomé BERA, head of the technical department of the Commune of Yaoundé II, for her sympathy and encouragement;
- the entire IRIC teaching staff and the 9th graduating class of the CA2D Master's program, for the skills and motivation they imparted
- all the students of the 9th promotion of the Master CA2D of the Institute of International Relations of Cameroon for the precious and unforgettable moments spent together during two years;
- my brothers and sister Fabrice, Ivan, Carole, Bertrand, Dylan, Junior, Evrard, Gustave, Eric for their encouragement and unfailing support
- my friends Philémon, Jordan, Auréole, Yoann, Fanny, Nancy, Fabrice, Freddy, Achille, Renaud, Cédric

SUMMARY

Generally speaking, decentralization refers to the transfer of powers and resources from central government to local or decentralized bodies. As such, it takes into account a system of administrative organization and management by which the State grants legal personality, financial autonomy and management to other territorial entities legally enshrined in the constitution. But beyond the choice of a particular form, which depends on geographical, historical and political factors, the implementation of decentralization offers at least three advantages: it contributes to more efficient public action, strengthens local development and improves equity through a better distribution of resources. The aim of this study is therefore to understand the extent to which decentralization can be a vector for local development, using the Yaoundé II Commune d'Arrondissement as a case study.

Key word : Decentralization - Local development - Commune

GENERAL INTRODUCTION

I. CONSTRUCTION OF THE OBJECT OF STUDY AND JUSTIFICATION OF THE SUBJECT

Since the 19th century, nation-states have been considered the main players in international relations. The twentieth century saw the emergence of new institutions of international scope[1] . These included the League of Nations, then the UN, UNESCO and the EEC, as well as a whole range of international intergovernmental organizations, non-governmental organizations (NGOs) and transnational corporations. Today, we know that if the 20th century was the time of the power of the Nation-State, this power has been weakened for several decades now by the processes of globalization, regionalization and decentralization affecting most States[2] . Local and regional authorities have thus become players in their own right on the international stage. The diversity of their activities is striking: decentralized cooperation, twinning, European programs, economic development initiatives, academic cooperation, humanitarian aid, advocacy, and so on.

This international action by local authorities may be encouraged by the central government's inability to satisfy the interests of local authorities on the national and international stage. It may also be motivated by political, cultural and economic particularities specific to each local government. This phenomenon is gaining in scope and recognition.

[1]Dario BATTISTELLA, Franck PETITEVILLE and Pascal VENNESSON , *Dictionnaire des relations internationales*, 3rd edition, Dalloz, 2012, p. 7

[2] Ibid.

As for the term "local development", it first appeared in France in 1965, promoted by a handful of pioneers and without any support from public authorities. It was a concept rooted in the Third World ideology of the late 60s, before decentralization became clearly established in most Western countries[3] . The concept took off in the 70s in rural areas. It was then exported to developing countries with the advent of structural adjustment programs, insofar as the aid policies promoted by donors in the South were directed towards local areas[4] .

Cameroon has thus been committed to this dynamic since it adopted decentralization as a mode of territorial administration through the Constitution of January 18, 1996[5] . Nevertheless, in Cameroon, as in most French-speaking African countries, the history of decentralization is highly complex. It began with colonization and continued through to the post-colonial period[6] . During both periods, the African commune, and Cameroon's commune in particular, grew out of the administrative district created by the colonizer for territorial management purposes, whose geographical contours it followed[7] . Despite the complexity of socio-historical considerations, the origins of decentralization, and more specifically of the communal movement in Cameroon, can be traced back to 1916, a period marked by the German colonial administration[8] . On this date, Germany, the great loser of the First World War, was forced to

[3] George GONTCHAROFF, Le développement local: petite généalogie historique et conceptuelle, in *Territoires*, Octobres 2002, p. 1-3.

[4] Bernard HUSSON, le développement local, in *Agridoc n°1*, available at: www.resacoop.org

[5] The preamble to the Constitution states that Cameroon is a decentralized unitary state, marking a major institutional and normative revolution at state level.

[6] Landry NGONO TSIMI, *L'autonomie administrative et financière des collectivités territoriales et décentralisées : l'exemple du Cameroun*, Doctoral thesis defended at the University of Paris-Est, 2010, p. 32

[7] *Ibid.*

[8] *Ibid*, p. 33

abandon its colonial possessions, which were henceforth administered by the League of Nations, in accordance with the Treaty of Versailles of June 28, 1919. This gave the British and French powers a mandate to administer Cameroon. The communal movement was thus marked by the influence of these two powers. Indeed, while the British power implemented the indirect rule system, the French colonial administration implemented direct administration[9] . Thus, in 1944, there were more than thirty communes in West Cameroon, a number which rose to 28 in 1967, then 24 again in 1967, then 24 in 1969, a number which remained unchanged until the adoption of the 1972 Constitution and the law on communal organization of the United Republic of Cameroon of December 5, 1974[10] . In East Cameroon, the first communes were created by the French Governor of Cameroon's decree of June 25, 1941, which created the mixed communes of Douala and Yaoundé[11] in the country's two largest conurbations. After the Second World War, a decree of August 21, 1952 created mixed rural communes and extended them to all subdivisions of the country. Three years later, French law no. 55-1489 of November 18, 1955 on municipal reorganization in Black African countries, with the exception of Senegal, introduced full-function communes and medium-function communes[12] . In 1967, the law of March 1 created the first "special regime" variants within the full-function communes, in the major cities of Douala, Yaoundé and Nkongsamba. These communes would later become urban communities, headed by government delegates appointed by the President of the Republic[13] . At the dawn of Reunification, Cameroon had 339 communes under the aegis of the Communal Law of

[9]*Ibid.*
[10] Landry NGONO TSIMI, *Ibid,* p. 36
[11]*Ibid.*
[12]*Ibid.*
[13]*Ibid.*

December 5, 1974[14] . The Constitutional Law of January 18, 1996 restored the Regions to the model of the pre-existing provinces, so that today there are some 384 decentralized territorial authorities in Cameroon, following the implementation of the provisions of the decentralization laws of July 22, 2004 and the numerous arrondissements created. Beyond these historical considerations, it is worth mentioning that, in terms of legislation, decentralization in Cameroon is now governed by the Code Général des Collectivités Territoriales, adopted on December 24, 2019. Institutionally, the State oversight of the decentralization process is currently ensured by the Ministry of Decentralization and Local Development, created by Decree no. 2018/449 of August 01, 2018, which succeeded the Ministry of Territorial Administration and Decentralization.

Following on from the above considerations, we understand that the current legal and socio-political contexts, marked by increased decentralization, are pushing local authorities to become major players in local development and international cooperation. It is with a view to putting this logic to the test of reality that we aim in this research to highlight the public action of the Commune of Yaoundé II. To this end, our work is entitled: ***"Local development and cooperation policies in Cameroonian communes tested by the strengthening of decentralization"***.

II. CONCEPTUAL CLARIFICATIONS

[14] *Ibid.*

The importance of conceptualization made Herbert BLUMER think that there can be no science without concepts[15] . In the same vein, Madeleine GRAWITZ asserts that a concept must be rigorously defined. For her, "*It is a universal and abstract mental representation, obtained by retaining the essential aspects of the object. Definition is the delimitation by its characteristics of the field of research*"[16] . These concepts must be defined according to the case under study[17] . With a view to building a solid research project, it is therefore imperative to proceed with an operational definition of the key concepts of our subject. We need to define: *"public policy"; "Local Development*"; "*Cooperation*"; *"Commune"; "Decentralization".*

A. Public policies

According to Pierre MULLER, the emergence of public policies was favored by the end of the feudal order and the emergence of the State in Europe between the 17th and 18th centuries.[18] .

As a discipline, the very first definition of public policy was given by Harold LASSWELL in 1936[19] . For him, public policy was a

[15]Herbert BLUMER, quoted by Howard BECKER in *Les ficelles du métier*, Paris, Guides Repères, La Découverte, 2002, p.180.

[16]Madeleine GRAWITZ,*Méthodes des sciences sociales,* Paris, 11th edition, Dalloz, 2001 , p.53.

[17] Howard BECKER, *op. cit.* p.198.

[18]He states: *"From the perspective we're interested in here, i.e. the emergence of public policy, two fundamental developments should be highlighted. The first is the monopolization by the king of a number of powers concerning taxation, currency, police and war. These regal functions were to form the basis of the modern state, while at the same time enabling the king to assert himself against the feudal lords. The second development is less visible but just as important. Following in Michel Foucault's footsteps, this involved the creation of a body of "governmental knowledge", i.e. all the "technologies" that would enable the State to govern territories and populations. This "governmentalization" changes the relationship between power and society, because the State now recognizes its legitimacy through its ability to produce order by implementing knowledge (such as statistics, for example) and effective measures (combating epidemics, organizing trade, etc.)* "Pierre MULLER, "L'analyse cognitive des politiques publiques : vers une sociologie politique de l'action publique", in *Revue française de science politique*, 50e année, n°2, 2000, pp. 189-208.

[19]Harold LASSWELL, *Who gets what, when and how*, Cleveland (Ohio), Meridian Books, 1936.

normative science, designed to determine which problems should be dealt with as a priority by political and administrative authorities: this was known as "*policy science"*. Long before this, however, *"cameral sciences" had* already developed in 18th-century Germany, with the aim of studying the collegial structure of administration. In the 19th century, this knowledge was structured and became *administrative science*, with the aim of describing administration and improving administrative practices (the State must achieve the well-being of society, and must equip itself with the right means to do so). At the end of the 19th century, administrative science declined in favor of administrative law: knowledge of the administration became legal and revolved around jurisprudence. Sociology also developed, and the State became an object of analysis. The real start of Public Policy was made in the United States at the beginning of the twentieth century, when the federal state was criticized and its powers greatly increased. Hence the need for analysis and knowledge with the aim of good administration, i.e. administration that is efficient (adapted to solving social problems) and responsive (focused on the well-being of citizens). Under the influence of Woodrow Wilson, a discipline called *"Public Administration"* was developed.

To define the concept of "*Public Policy"* today, we first need to look back at the basic concept of *"politics"*. From its English translation come three terms with distinct meanings: "*polity", "politics" and "policy"*. While the first term derives its etymological origin from the Greek word "*polies"* and alludes to the political organization of society, the second is roughly translated into French as *"politique",* and refers to the competition between participants in the political game to gain access to the legitimate exercise of power. Finally, the third term has the French equivalent of *"une politique", i.e. the set of measures taken by political authorities to resolve problems inherent in specific areas of social life*. Paquin argues that public policy is intimately linked to these three concepts: it represents a vast range of policies *(Policy)* that are only possible through *politics (politique)*, and which enable the political organization of society *(polity)*[20] .

Richard ROSE defines public policy as: *"a specific combination of laws, credit allocations, administrations and personnel directed towards a set of more or less clearly defined objectives"*[21] . Thomas DYE, for his part, states that public policy refers to what the State decides to do or not to do.

This concept also refers to interventions by an authority invested with public power and governmental legitimacy in a specific area of society or territory[22] .

Interventions can take three main forms: public policies convey content, translate into services and generate effects. They mobilize activities and work processes. They unfold through relationships with other social actors, whether collective or individual.

Public policy would be everything that government actors decide to do or not to do, actually do or do not do[23] . While it's relatively easy to list their actions, it's far less easy to identify their "non-actions":

---

[20]Stéphane PAQUIN, Luc BERNIER and Guy LACHAPELLE (eds.), *L'analyse des politiques publiques*, Les Presses de l'Université de Montréal, 2010, p. 8.

[21] Quoted by Daniel KLÜBER and Jacques de MAILLARD in *Analyser les politiques publiques*, Collection Politique en plus, Presses universitaires de Grenoble, p. 8.

[22]Madeleine GRAWITZ, Jean LECA and Jean Claude THOENIG, *Les Politiques publiques*, Tome 4, Paris, PUF, 1985

[23] Yves MENY and Jean-Claude THOENIG, *Politiques publiques*, Paris, PUF, 1989

what they refuse to do or avoid doing. It's often difficult to identify a program immediately and in its entirety. When a policy is proclaimed in a sector, laws and resources often outline an incomplete perimeter of resources, and hardly specify the conditions of their concrete application.

A public policy takes the form of a specific program carried out by a government authority[24] . This authority acts in two ways: through materially identifiable practices (controls, construction and maintenance of infrastructures, allocation of financial subsidies, dispensation of care, etc.) and through immaterial practices (institutional communication campaigns, speeches, propagation of norms and cognitive frameworks).
Every public policy implicitly or explicitly conveys a segmentation of audiences. It targets individuals. It postulates causalities between actions and results[25] .

Beyond the competition for power, the analysis of public policy highlights the all-too-frequently hidden face of government work. This work comprises a number of phases, including setting an agenda, taking or failing to take account of the demands or interests expressed by the social body, drawing up action programs, and formulating and legitimizing solutions to problems that have become public[26] .

---

[24] Pierre MULLER, *Les Politiques publiques*, Paris, PUF, Coll. "Que sais-je", 2009, 8th ed.
[25] Jean-Claude THOENIG, *Ibid.*
[26] Charles O. JONES, *An introduction to the study of public policy*, Belmont (California), Duxbury Press, 1970

From a sociological point of view, public action refers to the programmed set of goals, values and practices built at the heart of social interactions rather than at the top of the State, and is therefore fragmented, complex and flexible. This is the definition we will use throughout this work.

B. Local development

In France, the concept of local development emerged in the 1970s in rural areas: *"it was born in response to the risk of economic, demographic and social desertification of regions disadvantaged by economic change and the development of industrial and urban centers. Indeed, it was in these regions that local players first felt the need to define a form of development other than that of economic growth or planned development"*[27] .

As far as developing countries are concerned, during the period of structural adjustment, donors, drawing on the experience of Western economies in local development, will seek to articulate the logics of States with those of the market and territories. In other words, donor aid policies aimed at developing countries in the South, like public development policies, were to focus on local areas.

The aim of this reorientation, which took place in the 1980s, was to enable the various development policies to be based on local will. Little by little, we moved from *"a view of a 'passive' local area dependent on the goodwill of external decision-making centers, to*

[27] Yves AUTON, "Etudes internet et développement local, première partie : le développement local", 2000 , available at www.admiroutes.asso.fr.

a local area perceived as being able to generate its own development dynamics by drawing on its capacity for initiative and organization"[28] .

For Bernard HUSSON, local development is a decision-making process that involves a large part of the population in the strategic orientation of the territory, with a view to satisfying their needs. According to HUSSON, this type of development is not at odds with state-led development. However, the new interest in local development and hence decentralization shows that these processes are very similar. And yet, they are two distinct processes, which is why we need to make clear the difference between local and communal development.

Bernard HUSSON believes that local development focuses on actors rather than infrastructures, on networks rather than institutions, to give people and groups a decision-making role in the actions they undertake. Communal development, on the other hand, is implemented by communes, organizations with institutional legitimacy that operate in a circumscribed territory and within a field of competence delimited by law. The decisions they take are binding on all[29] .

Crucially, he adds that *"failure to take account of the difference between local development and communal development risks focusing support on emerging local authorities and marginalizing*

---

[28] Bernard HUSSON, *op cit.*

[29] Bernard HUSSON, "La coopération Décentralisée : légitimer un espace publique local au Sud et à l'Est", 2000, available at www.capcoopération.org, last consultation date June 13, 2020.

interesting programs run by private organizations and individuals".[30]

For the purposes of this work, we will adopt Bernard HUSSON's definition of local development as "a decision-making process that involves a large part of the population in the strategic orientation of the territory, with a view to satisfying their needs".

C. Cooperation

Cooperation - its emergence, duration, purpose, modalities and limits - is a major issue in international relations, and the focus of numerous theoretical, methodological and empirical debates. Cooperation is essential for preserving peace, enabling economic or diplomatic exchanges, or ensuring the durability of a certain international order. Inspired by game theory and the main research traditions in the study of international relations[31] (notably realism, liberalism and constructivism), several theories offer models for better understanding the logics of cooperation (and conflict)[32] . Analysts and practitioners of realist inspiration generally believe that the prospects for cooperation are few and fragile, and that international law and institutions cannot foster it[33] . The various variants of liberal conceptions (functionalism, regional integration, neo-liberal institutionalism) reject this point of view and consider, on the contrary, that cooperation is possible in many ways and, in

[30]*Ibid.*
[31] Jervis SCOTT, Realism, game theories and cooperation,inWorlds *Politics*, XL (3), 1988, pp. 317-349
[32] Dario BATISTELLA, Frank PETITEVILLE and Pascal VENNESSON, *op.cit*, p. 78.
[33]Jervis SCOTT, *Realism, Neoliberalism and cooperation understanding the debate in International Security*, 24(1), 1999, p. 42 - 63

particular, that international institutions are conducive to it[34] . The question is: what are the factors behind the adoption of cooperative behavior?

According to Robert AXELROD, the first process concerns the ways in which the actors' common interests converge, and relates to the benefits that states derive from their cooperative or non-cooperative policies[35] . These are the benefits and penalties associated with the situations identified by the prisoner's dilemma: mutual cooperation, cooperation by one state and defection by another, or mutual defection. According to AXELROD, in international relations, trust, loyalty and altruism are not dispensable to cooperation[36] . The durability of relationships, the repetition of strategic interactions between the players concerned, may be enough to ensure forms of cooperation. The second process concerns the number of players and its influence on the way the game is played, in particular the hegemony of one player imposing his rules[37] . Finally, the third process concerns institutional mechanisms, i.e. the characteristics of strategic interaction, of the "international game", at a given moment, which influence the propensity of players to cooperate. Institutions can, for example, promote transparency and enable players to identify those who have cooperated or those who have defected in the past[38] .

---

[34]Jervis SCOTT, Ibid.

[35] Robert AXELROD, *Donnant donnant, Théorie du comportement*, Paris, O. Jacob, 1ère edition, 1992, p. 50

[36] Robert AXELROD, *Ibid*, p. 37.

[37]Robert AXELROD, *Ibid*.

[38]Robert AXELROD, *Ibid*.

From a constructivist perspective, a sense of common belonging, mutual respect, similar identities and the acquisition of a reputation for reliability all contribute to cooperation and its institutionalization. In addition, transnational factors can play a significant role in the establishment of a "security community" guaranteeing the development of trust beyond a minimal agreement: the refusal to resort to force in the settlement of disputes[39] . From this perspective, external constraint is less important than the collective feeling that creates a pluralist community, of which the North Atlantic Treaty Organization, for example, would be the emanation[40] . Transnational identities can therefore contribute to the creation of a community, even in the context of international anarchy[41] .

Finally, the third debate on cooperation in international relations concerns the role of institutions. What influence do international institutions and regimes have on cooperation? From the perspective of neoliberal institutionalism, institutions stabilize the stakes, promote transparency, make the future more predictable and thus enable states to cooperate[42] . Institutions and regimes reduce the uncertainties involved in assessing the preferences and political choices of potential partners. These institutions facilitate a sustainable exchange of information, clarifying intentions through consultation procedures[43] . All in all, transaction costs are reduced.

---

[39] Karl DEUTSCH,*Seminar onPolitical community and the North Atlantic: international organization in the light of historical experience*,1957.

[40]Dario BATISTELLA, Frank PETITEVILLE and Pascal VENNESSON, *Ibid*, p. 82.

[41] Bruce CRONIN, *Community under anarchy. Transnational identity and the evolution of cooperation*, New York, Columbia University Press, 1999.

[42]Dario BATISTELLA, Frank PETITEVILLE and Pascal VENNESSON, *Ibid*, p. 83

[43]*Ibid.*

The benefits derived from institutions are all the more important given that security issues are as much military as they are economic or environmental, and that effective collective action at international level must bring together a wide variety of players with heterogeneous natures, capabilities and interests[44] . This process is particularly significant when it comes to alliances. In a field as sensitive as security, for example, the initial political investment is heavy, and the players will want to preserve the framework they have created. An institution such as the North Atlantic Treaty Organization, for example, will stay because the cost of its demise will be greater than the cost of adapting it[45] . For realists, on the other hand, the institutionalization of cooperation in no way alters the behavior of states, which respond to the logic of their interests and their power[46] . They recognize their importance as an instrument of states' external action, but stress that they are all the more present when cooperation is already high: states create institutions if they want to achieve the objectives that these institutions will help them to attain[47] . Institutions and regimes are therefore not causes, but effects. Ultimately, cooperation remains one of the most important and challenging enigmas of international relations. For the purposes of our research, we will adopt the realist understanding of the concept of cooperation as a political process by which states pool the institutional means necessary to achieve their goals.

D. Municipality

---

[44] *Ibid.*
[45] *Ibid.*
[46] *Ibid.*
[47] *Ibid.*

According to EsohElame, the Commune is a small delimited portion of a country's territory. It is a territorial district that may correspond to a town with its villages, or to a group of villages. In formal terms, especially in several Western countries, it is the smallest administrative subdivision[48] .

E. Decentralization

Decentralization is an administrative and political management technique by which the State creates infra-state entities, to which it transfers the competences and resources necessary for their development. Decentralization can take several forms (technical and territorial). In the context of this research project, we will focus on territorial decentralization. According to Article 2 of the Law on Decentralization, decentralization consists of "the transfer from the State to local authorities of specific powers and appropriate resources". Territorial decentralization thus aims to bring decision-making closer to action, by making people more responsible for their environment.

According to George LUTZ and Wolf LINDER, it's clear that successful decentralization requires more than just good political institutions. It is also essential to improve overall governance at local level. This means effective participation of local populations and their inclusion in decision-making processes. The effective

[48]Esoh ELAME, *Decentralized Cooperation Policy and Engineering Course*, IRIC, 2020.

inclusion of all stakeholders at local level is crucial to the success of local development[49] .

III. SUBJECT DELIMITATION

Michel BEAUD, advises us to limit the study by giving it a field of definition[50] . This approach ensures the feasibility of the research. It is with this in mind that we proceed with the spatial (A) and temporal (B) delimitation of our field of investigation.

A. SPATIAL DELIMITATION

The spatial delimitation of our subject will be limited to the city of Yaoundé, more precisely to its second arrondissement, the area of jurisdiction of the Commune that is the subject of our research.

B. TIME DELIMITATION

Our timeframe runs from 2004, the year in which the first decentralization laws were enacted in Cameroon, to 2019.

IV. INTEREST OF RESEARCH

Gaston BACHELARD states that "*the scientific spirit is essentially a rectification of knowledge, a broadening of the frameworks of knowledge"*[51] . All research must therefore make a contribution to the

[49] George LUTZ and Wolf LINDER, *Structures traditionnelles dans la gouvernance locale pour le développement local,* University of Berne, Institute of Political Science, Switzerland, 2004.

[50]Michel BEAUD, *L'art de la thèse*, Paris, La Découverte, revised edition, 2006 p. 27.

[51]Gaston BACHELARD, *Le nouvel esprit scientifique*, Paris, Les Presses universitaires de France, 10th edition, 1968, p. 131.

advancement of science, and this contribution is generally formulated in terms of interest. As the present research is fully in line with this logic, its interest is both scientific (A) and practical (B).

A. HEURISTIC INTEREST

For Lawrence OLIVIER, Guy BEDARD and Julie FERRON, the interest and relevance of research is first and foremost theoretical[52] . From this perspective, the particularity of this work lies first and foremost in its interdisciplinary approach. Indeed, while highlighting how decentralization and local development are becoming issues of international cooperation, it mobilizes a theoretical corpus specific to the analysis of public policies, enabling us to highlight their design and implementation at local level.

B. PRACTICAL INTEREST

BOURDIEU postulates that "*it is possible to act on the social world by intervening on the knowledge that agents have of it*"[53] . In keeping with this line of thought, this research, while diagnosing the level of appropriation and implementation of the concepts of decentralization and local development in Cameroon, questions the efficiency of the institutions in charge of these tasks. This research also aims to make a modest contribution to decision-makers, with a view to improving or reorienting these various local policies. Indeed, insofar as local

---

[52]Laurence OLIVIER, Guy BEDARD and Julie FERRON, *L'élaboration d'une problématique de recherche: sources, outils et méthodes*. L'Harmattan, Logique sociale, 2005, p. 28.

[53] Pierre BOURDIEU, *Propos sur le champ politique*, Presse Universitaire de Lyon, Lyon, 2000, p. 17

authorities or decentralized territorial communities are now considered to be the foundation of all development on a national scale, this research is intended as a plea for local authorities to be strengthened by the State in the local development process.

Having clarified the scientific and practical contributions of our research, it is now time to review the related literature, and raise the problematic aspect.

V. LITERATURE REVIEW

For OLIVIER, BEDARD and FERRON, the central aim of the literature review is to problematize what has been said in the existing literature[54] . The search for works relating to our object of study led us to select the most relevant documents from the abundant literature.

In this section, we begin with the work of Santiago BETANCUR RAMIREZ[55] .The author examines the factors that have led to the emergence of local governments as players on the international stage, asking how this trend has evolved over time and what governance issues have arisen in a sphere involving entities at different levels (local authorities, States, supranational organizations). More specifically, what are the characteristics of this practice in France, particularly in its relations

---

[54]*Cf.* Laurence OLIVIER, Guy BEDARD, Julie FERRON, *op. cit.* p. 75.

[55]Santiago BETANCUR RAMIREZ, *What role for local governments on the international scene? L'action internationale des collectivités locales entre la France et l'Amérique latine,* dissertation defended publicly for the Master II in Public Affairs,ENA, Paris, 2018, 171p.

with Latin America? To answer these questions, and to gain a better understanding of the phenomenon of international action by decentralized local authorities, the author first looks at the conditions under which local governments have emerged on the international scene. It then focuses on how these practices have developed and become established. While questioning the multi-level modes of governance that have become established, the author transposes all these reflections to the French case, where he addresses the evolution of this phenomenon and decomposes the international action of French local authorities in Latin America, establishing a panorama of this dynamic in this region. Finally, he examines the reticular relationships of Euro-Latin American local governments through a case study.

Then we were struck by the relevance of Yannick Félix PEGUI's work[56] . The author examines the decentralization process in Cameroon and the changes it has undergone since the 2000s. To this end, he explores the various legal and institutional reforms granting new prerogatives to decentralized local authorities. Using the Commune d'Arrondissement of Yaoundé II as a framework for analysis, the author makes the empirical observation that decentralization is beginning to yield remarkable results, which should be capitalized on. The author also notes the important role played by decentralization support institutions, and postulates that this should be encouraged and strengthened in the interests of greater support for decentralized territorial authorities, insofar as most of them lack the resources to fully exercise some of the powers transferred to them by the

[56] Yannick Félix PEGUI, *Décentralisation et fonctionnement des Communes au Cameroun. Cas de la Commune d'Arrondissement de Yaounde II*, dissertation defended publicly for the award of the Master II in Economic Sciences, University of Yaoundé II, Yaoundé, 2012,

State. It also recommends a pragmatic application of the principle of subsidiarity, and the creation of a genuine local civil service.

We were then drawn to the work of LANDRY MEPUI ABAH[57] . The author begins with a cold, hard-hitting diagnosis of the development process underway in Cameroon. He argues that decentralized governance, which was supposed to drive local development, is failing to do so. According to the author, the politico-administrative machinery has stalled, mainly due to a host of seemingly insoluble problems: a lack of mastery of the nuts and bolts of decentralization and decentralized governance on the part of all stakeholders, limited support for local authorities, laws and texts that do not really respond to the needs of local authorities, and a lack of understanding of the major issues at stake in local development on the part of certain stakeholders. As a consequence of this state of affairs, the author highlights administrative and managerial excesses such as inertia, laxity, corruption and misappropriation of local public funds, which keep communities in a precarious situation and thus compromise Cameroon's emergence by 2035. Yet the author believes that decentralized governance can be a lever for achieving Cameroon's development objectives. To this end, he proposes a roadmap that emphasizes the primordial role of the State in driving development, while inviting all players involved in decentralized governance to play their part.

We were also intrigued by the article by MURIEL SAME EKOBO and OLIVIER IYEBI MANDJEK[58] . This article sets out to analyze the decentralization process in Cameroon by adopting the institutional and

[57] Landry MEPUI ABAH, *Dynamiques des politiques décentralisées au Cameroun, Analyse des enjeux et défis sociopolitiques économiques et culturels, L'Harmattan,* Paris, 2012, 115p.

[58] Muriel SAME EKOBO and Olivier IYEBI MANDJEK, "Gouvernance territoriale et action publique au Cameroun", in *Enjeux*, n°45-46, July 2012, p. 9

transactional political analyzer of *"governance"*, which operates as a *"new mode of territorial coordination". In* so doing, the aim is to proceed with the theoretical and empirical implementation of a critical political geography of territorialized public action, capable of *"highlighting the shortcomings of public action in the context of decentralization"* and underlining the obstacles weighing on *"a process of (re)legitimizing the representation of different public actors and the participation of civil society".*

We then turned our attention to the article by GERARD PEKASSA NDAM[59] . The author makes a critical assessment of *"the place of the administration"* in *"decentralization policies"* and *"local governance".* He then proceeds with a legal-institutional and legal-political reading that highlights the limits of a *"synergy of actions between the center and the periphery".* This enables the author to show how decentralization and local governance policies in Cameroon remain relativized in their capacity to consolidate democratic participation, which still leaves an important part to a *"hegemonic place of state administration but marginal in local administration".*

Our attention was then drawn to MATHIEU *MEBENGA*'s article[60] . The author focuses on the place of *"financial decentralization"* as *"the common heritage of the nation".* For this reason, *"taxation",* as one of the *"sources of funding for decentralization", is the* subject of an analysis that highlights its *"exact function"* in the *"context of decentralization". Rightly so*, the author examines *"relations between the State and decentralized*

[59] Gérard PEKASSA NDAM, "La place de l'administration dans les politiques de décentralisation et de la gouvernance locale au Cameroun", in *Enjeux*, n° 45-46, July 2012, p. 14

[60] Mathieu MEBENGA, "La fiscalité dans la gouvernance décentralisée au Cameroun", in *Enjeux*, n° 45-46, July 2012, p. 17

local authorities", and identifies *"taxation as a decentralized financial governance technique"*. The author also questions the paradoxical use of taxation as an "economic development strategy".

VI. PROBLEMS

Generally speaking, the problematic is defined as "the *search for or identification of what poses a problem"*[61] . For CHENIER, it refers precisely to *"the set of elements forming a problem, to the structure of information whose interrelationship generates in a researcher a deviation resulting in an effect of surprise or questioning sufficiently stimulating to motivate him to carry out research"*[62] . In short, it can be likened to the theoretical approach or perspective we decide to adopt to address the initial question. It represents a pivotal stage between rupture and construction[63] . After reading the documents included in our literature review, we realize the complexity of the current context, marked by a strengthening of decentralization. This context has led communes to play a major role in the appropriation of local development policies. As a result, understanding the public action of the Commune d'Arrondissement de Yaounde II in terms of local development and international cooperation leads us to an in-depth questioning, hence the formulation of our problematic composed of a main question (A) and two secondary questions (B).

A. MAIN QUESTION

[61]Lawrence OLIVIER, Guy BEDARD, Julie FERRON, *op. cit.* p.24

[62]*Ibid*, p.11.

[63] Raymond QUIVY, Luc Van CAMPENHOUDT, *Manuel de recherche en sciences sociales*, Paris, Dunod, 2nd edn, 1995, pp.98-100.

In the current context of increased decentralization, can we talk about the implementation of local development policies and international cooperation in the Yaounde II Commune d'Arrondissement?

In addition to this first question, we need to ask a number of secondary questions to better understand the problematic aspects of our subject.

B. SECONDARY ISSUES

In fact, these two secondary questions flow from the central question.

Secondary question 1:

What are the characteristics of the process of strengthening decentralization in Cameroon?

Secondary question 2:

Does the Commune d'Arrondissement of Yaounde II have departments responsible (respectively) for local development and international cooperation, and if so, what are they doing?

Following these questions, which will be the focus of our research, we propose a number of hypotheses to guide our progress in this heuristic process.

VII. HYPOTHESES

A hypothesis can be defined as a provisional answer to a question. According to QUIVY and CAMPENHOUDT, it must be falsifiable[64] . It serves to bridge the gap between theoretical reflection and verification. In this respect, according to Gordon Mace, it is a precursor to operationalization, since it concretizes the abstract relation stated at the end of the problem formulation, i.e., it transforms the theoretical concepts of the specific question into operative concepts. Thus, as a provisional answer to the main question, we propose a main hypothesis:

A. Main assumption

Looking at some of the actions carried out by specialized departments, it is indeed possible to affirm that the Yaoundé II Commune d'Arrondissement is in line with the dynamics of strengthening decentralization by implementing local development policies and actions within the framework of international cooperation.

B. Secondary hypotheses

There are two of them.

Secondary hypothesis 1:

The dynamics surrounding the strengthening of the decentralization process in Cameroon are characterized by a normative and

[64]Raymond QUIVY, Luc Van CAMPENHOUDT,*Ibid*.

institutional development of the socio-political environment according to a chronological logic.

Secondary hypothesis 2:

The Commune d'Arrondissement of Yaoundé II has a local cooperation and development department, which, under the supervision of the Mayor, implements targeted actions in the areas for which it is responsible.

VIII. THEORETICAL FRAMEWORK

A theory is a principle or rule on which rational knowledge is based, as well as a body of principles that can be used to explain and understand a fact. It represents the result of an ordered process of reflection that follows the observation of facts according to a set of hypotheses. Reference to a theory is made either because it validates the facts, or because it contradicts them. Reference to a particular theory must be justified, as must the way in which it explains economic or social facts. For the purposes of this research, we will focus on the cognitive approach to public policy (A) and sociological institutionalism (B).

A. THE COGNITIVE APPROACH TO PUBLIC POLICY

The cognitive approach to public action refers to a wide variety of works. From the perspective developed here, the cognitive approach to public policy should not be considered as an approach based on ideas. As we shall see later on, this is a point of debate among the various authors.

Some, like Eve FOUILLEUX for example, explicitly following in the footsteps of Péter HALL[65] , value an idea-centred approach. This approach seems to us to be both risky from a scientific point of view. It's risky because, despite all the precautions we can take, it's virtually impossible to avoid the trap of being trapped in a sterile debate between ideas and interests. On the contrary, we must reaffirm that the cognitive approach to public policy is not opposed to an approach based on interests and institutions, since it considers that the interests at stake in public policy are only expressed through the production of frameworks for interpreting the world. In this sense, the approach proposed here is quite distinct from that of EDELMAN[66] , which establishes the importance of symbolic and rhetorical elements in the determination and use of policies, and seeks to go beyond this admittedly unquestionable observation of the symbolic nature of public action. For the same reason, while readily acknowledging the proximity between these different works, she does not identify Péter HALL's approach in terms of paradigm[67] on the one hand, the concept of referential is similar to that of paradigm, insofar as in both cases it is possible to identify normal phases, i.e. periods in which a certain framework of interpretation is established.i.e. periods when a certain framework for interpreting the world is more or less accepted and recognized as true by a majority of actors, and phases of crisis during which a number of anomalies appear, testifying to the growing inability of the paradigm or frame of reference to account for reality. On the other hand, what distinguishes the paradigm from the frame of reference concerns the conditions under which they are invalidated: whereas a

---

[65] Peter HALL, Policy Paradigms Social Learning and the State, *Comparative Politics* ,n°25 (3), 1993, p. 275-296

[66] M. EDELMAN, *The Symbolic Uses of Politics*, Urbana, University of Illinois Press, 1976

[67] For a paradigm-based approach, see HALL Policy Paradigms. Art cité and Y. SUREL, *Idées intérêts et institutions dans analyse des politiques publiques*, Pouvoirs, 1998, p.87

paradigm will be invalidated in fine by the test of experimental verification, the same is obviously not true of invalidation by a frame of reference, which will be based on a transformation of the beliefs of the actors concerned. But the cognitive analysis of public policy also differs from cognitive sociology, even if it adopts many of its achievements, insofar as it does not adopt a methodological individualist perspective, as advocated in the work of Raymond Boudon. In fact, the cognitive analysis of public policy inherits a conception according to which, even if cognitive matrices are indeed produced by the interaction of individual actors, they tend to become autonomous in relation to their construction process, and to impose themselves on actors as dominant models for interpreting the world. The cognitive analysis of public policy is therefore based on a moderate constructivism[68] , not everything is constructed. This underlines the irreducibility of the political function in relation to the processes of expressing interests and, more generally, the processes of cognition as perceived at the level of individuals. Having made these preliminary remarks, we can say that the starting point for the cognitive analysis of public policy is the observation that public policy does not serve, or at least not only serves, to solve problems. Of course, this is not to say that public policy has nothing to do with solving public problems, the existence of which is unfortunately undeniable. The point is to realize that, contrary to what political leaders would have us believe and what some public policy analysts sometimes suggest, the relationship between public action and public problems is far more complex than the common idea that policies only serve to solve problems suggests. As Jean Leca clearly shows[69] , the dilemma facing government today is to be *responsive*

[68] P .BERGER, T. LUCKMANN, *La construction sociale de la réalité*, Paris, Méridiens-Klincksieck, 1986

[69] Jean LECA, La gouvernance de la France sous la Cinquième République Une perspective de sociologie comparative in F. D'ARCY, L. ROUBAN (eds.) *De la Vé République à l'Europe*, Paris, Presses de Sciences Pô, 1996.

(aware of the problems and demands of the population), *accountable* (likely to be held to account for its actions, which presupposes that it knows what it is doing and what the results are) and *problemsolving* (capable of solving problems).

Within the framework of this research, the analysis of public policy cognitives will enable us to grasp the different dynamics that have led to the appropriation of the decentralization process by the State of Cameroon.

B. SOCIOLOGICAL NEOINSTITUTIONALISM

Parallel to these developments in political science, a neo-institutionalism has developed in sociology. Like other schools of thought, it is fraught with internal debate. However, its proponents have developed a series of theories that should be of considerable interest to political science researchers. What we call sociological institutionalism has its origins in organization theory. This movement dates back to the late 1970s, when certain sociologists began to challenge the traditional distinction between the sphere of the social world deemed to reflect an abstract rationality of ends and means (of the bureaucratic type), and spheres influenced by a varied set of practices associated with culture. Since Max Weber, many sociologists have viewed the bureaucratic structures that dominate the modern world - whether in ministries, companies, schools, interest groups, etc. - as the product of an intense effort to develop ever more efficient structures designed to accomplish the formal tasks associated with these organizations. They believed that the organizational form of these structures was virtually the same because of the rationality or efficiency inherent in these forms and necessary to fulfill these tasks[70] . Culture appeared to them as something quite different. Against this trend, neo-institutionalists began to argue that many of the institutional forms and procedures used by modern organizations were not adopted simply because they were the most efficient with regard to the tasks at hand, as implied by the notion of transcendent "rationality". On the contrary, they argued, these forms and procedures had to be considered cultural practices, comparable to the myths and ceremonies developed by many societies, and were therefore incorporated into organizations, not necessarily because they increased their abstract efficiency (in terms of ends and means), but because of the same kind of transmission process that gives rise to cultural practices in generai. Thus, even the most apparently bureaucratic practice had to be explained in terms of this culturalist grid[71] .

Given their perspective, institutionalist sociologists generally choose a problematic that seeks to explain why organizations adopt a given set of institutional forms, procedures or symbols, with an emphasis on the diffusion of these practices. They seek, for example, to explain the striking similarities, in terms of organizational form and practice, between Ministries of Education around the world, whatever the differences in context, or between companies belonging to different industrial sectors, whatever the product they manufacture. Frank DOBBIN uses this

[70] For a more detailed presentation, cf. F. DOBBIN, "Cultural Models of Organization. The Social Construction of Rational Organizing Principles", in D. CRANE (ed.), *The Sociology of Culture*, Oxford, Blackwell, 1994, pp. 117-153.

[71] The first to break new ground in this field were Stanford sociologists. Cf. J.W. Meyer, B. Rowan, "Institutionalized Organizations. Formal Structure as Myth and Ceremony", *American Journal of Sociology*, 83, 1977, pp. 340-363; J.W. MEYER, W.R. Scott, *Organizational Environments. Ritual and Rationality*, Beverly Hills, Sage, 1983. For a very good overview, cf. P. DIMAGGIO, W.W. Powell, "Introduction", in W.W. POWELL, P. DIMAGGIO (eds), *The New Institutionalism in Organizational Analysis*, Chicago, University of Chicago Press, 1991, pp. 1-40.

approach to show how culturally determined conceptions of the state and the market conditioned railway policy in France and the USA in the 19th century[72] . John W. MEYER and W. Richard SCOTT use it to explain the proliferation of training programs in American companies[73] . Others apply it to explain institutional isomorphisms in the Far East, and the relatively easy spread of production techniques from this area throughout the world[74] . Neil FLIGSTEIN uses it to explain the diversification of American industry, and Yasemin SOYSAL to explain current immigration policy in Europe and America[75] . Three features of sociological institutionalism give it a certain originality compared to other varieties of "neo-institutionalism". Firstly, theorists of this school tend to define institutions much more comprehensively than political science researchers, so as to include not just formal rules, procedures or norms, but the symbol systems, cognitive schemas and moral models that provide the "frames of meaning" guiding human action[76] . This position has two important consequences.

Firstly, it breaks down the conceptual dichotomy between "institutions" and "culture", as these two notions come to interpenetrate. As a result, this approach undermines the distinction that many political scientists like to draw between "institutional explanations", which consider institutions as the rules or procedures instituted by organizations, and "cultural

---

[72] F. DOBBIN, *Forging Industrial Policy*, Cambridge, Cambridge University Press, 1994.

[73] W. R. SCOTT, J.W. MEYER et al, *Institutional Environments and Organizations*, Thousand Oaks, Sage, 1994, chapters 11 and 12.

[74] M. ORRU et al, "Organizational Isomorphism in East Asia", in W.W. Powell, P. DIMAGGIO (eds), op. cit, pp. 361-389 and R. E. COLE, *Strategies for Industry: Small-Group Activities in American, Japanese and Swedish Industry*, Berkeley, University of California Press, 1989.

[75] N. FLIGSTEIN, *The Transformation of Corporate Control*, Cambridge, Harvard University Press, 1990; Y. SOYSAL, *Limits of Citizenship*, Chicago, University of Chicago Press, 1994.

[76] J. L. CAMPBELL, *"Institutional Analysis and the Role of Ideas in Political Economy"*, paper presented at the Seminar on the State and Capitalism since 1800, Harvard, 1995, and W.R. SCOTT, *"Institutions and Organizations: Towards a Theoretical Synthesis"*, in W. R. SCOTT, J.W. MEYER et al, Institutional Environments..., op. cit. 55-80.

explanations", which refer to culture defined as a set of shared attitudes, values and approaches to problems[77] . Secondly, this approach tends to redefine "culture" as synonymous with "institutions"[78] . In this respect, it reflects a "cognitivist turn" within sociology itself, moving away from conceptions that associate culture with norms, affective attitudes and values, towards a conception that sees culture as a network of habits, symbols and scripts that provide patterns of behavior[79] .

Neo-institutionalists in sociology are also distinguished by their way of looking at the relationship between institutions and individual action, which is a consequence of the "culturalist approach" roughly described above, but which develops certain particular nuances. An earlier school of sociological analysis solved the problem of the relationship between institutions and action by associating institutions with "roles" to which prescriptive "norms of behavior" were attached. According to this view, individuals who are socialized into particular roles internalize the norms associated with these roles, and it is in this way that institutions are supposed to influence behavior. We might refer to this conception as the "normative dimension" of institutional impact. Although some continue to use such conceptions, many theorists now insist on what we might call the "cognitive dimension" of the impact of institutions. In other words, they emphasize how institutions influence behavior by providing cognitive schemas, categories and models that are indispensable to action, one of the main reasons being that without them, it would be impossible to

[77] G. ALMOND, S. VERBA, *The Civic Culture*, Boston, Little Brown, 1963 and P. A. HALL, *Governing the Economy*, op. cit. chapter 1.

[78] Cf. L. ZUCKER, "The Role of Institutionalization in Cultural Persistence", in W.W. POWELL, P. DIMAGGIO (eds), *The New Institutionalism in Organizational Analysis*, op. cit, pp. 83-107; J.W. MEYER et al, "Ontology and Rationalization in the Western Cultural Account", in J.W. MEYER, W. R. SCOTT et al, *Institutional Environments and Organizations*, op. cit.

[79]Cf A. SWIDLER, "Culture in Action: Symbols and Strategies", *American Sociological Review*, 51, 1986, pp. 273-286 and J. MARCH, J. P. OLSEN, *Rediscovering Institutions*, op. cit., chap. 3.

interpret the world and the behavior of other actors[80] . Institutions influence behavior not simply by specifying what to do, but also what one can imagine doing in a given context. Here, we can see the influence of social constructivism on neo-institutionalism in sociology. In many cases, institutions are supposed to provide the very conditions for the attribution of meaning in social life. It follows that institutions influence not only the strategic calculations of individuals, as theorists of the rational choice school maintain, but also their most fundamental preferences. The identity and self-image of social actors are themselves supposed to be constituted from the institutional forms, images and signs provided by social life[81] .

As a result, many institutionalists insist on the highly interactive nature of the relationship between institutions and individual action, a relationship in which each pole constitutes the other. When they act in accordance with a social convention, individuals simultaneously constitute themselves as social actors, i.e. they undertake actions endowed with social significance and reinforce the convention they obey. A fundamental corollary of this view is the idea that action is closely linked to interpretation. Thus, theorists of sociological institutionalism maintain that, when confronted with a situation, the individual must find a way to identify it as well as to react to it, and the scenarios or models inherent in the world of the institution provide the means to accomplish both tasks, often relatively simultaneously. The relationship between the individual and the institution is thus based on a kind of "practical reasoning" whereby, to

[80] P. DIMAGGIO, W.W. POWELL, *op. cit.*

[81] P. BERGER, Th. LUCKMANN, *The Social Construction of Reality*, New York, Anchor, 1966 and its more recent application to political science by A. WENDT, "The Agent- Structure Problem in International Relations Theory", inInternational *Organization*, 41(3), Summer 1987, pp. 335-370.

develop a course of action, the individual uses the available institutional models while at the same time shaping them[82] .

Nothing in this suggests that individuals are not endowed with intentions, or are irrational. However, theorists of sociological institutionalism emphasize that what an individual tends to regard as "rational action" is an object that is itself socially constituted, and they conceptualize the goals an actor sets for himself according to a much broader grid than other theorists. While rational choice theorists often postulate a universe of individuals or organizations seeking to maximize their material well-being, sociologists describe a universe of individuals or organizations seeking to define or express their identity in socially appropriate ways. Finally, neo-institutionalists in sociology are distinguished by their approach to the problem of how to explain the emergence and modification of institutional practices. As we have seen, many theorists of rational choice institutionalism explain the development of an institution by reference to the efficiency with which it serves the material ends of those who accept it. In contrast, sociological institutionalists argue that organizations often adopt a new institutional practice less because it increases their membership. In other words, organizations adopt particular institutional forms or practices because they have a widely recognized value in a wider cultural environment. In some cases, these practices may be aberrant in relation to the fulfillment of the organization's official objectives. John L. Campbell puts it well when he speaks of a "logic of social propriety" as opposed to an "instrumental logic"[83] .

---

[82] P. DIMAGGIO, W.W. POWELL, "Introduction", in P.J. DIMAGGIO, W.W. POWELL. POWELL, *The New Institutionalism...*, op. cit. p. 22-24 and the essays by L. ZUCKER and R. JEPPERSON in the same volume.

[83] J.L. CAMPBELL, "Institutional Analysis and the Role of Ideas in Political Economy", quoted in J. March, J.P. Olsen, *Rediscovering Institutions*, op. cit., chap. 2.

Thus, in contrast to theorists who explain the diversification of American companies in the Fifties and Sixties as a functional reaction to economic or technological requirements, Neil FLIGSTEIN argues that entrepreneurs made this choice because of the value that came to be attached to this notion in the many professional forums in which they participated, and because this choice provided an endorsement of their social role and worldview[84] . Similarly, Yasemin SOYSAL argues that the immigration policy adopted by many states was pursued, not because it was the most functional for each state, but because the new conception of human rights proclaimed by international regimes made this policy appear appropriate, while others appeared illegitimate in the eyes of national authorities[85] . The fundamental question, from this point of view, is obviously what confers "legitimacy" on certain institutional arrangements rather than others. In the final analysis, this question implies a reflection on the sources of cultural authority. In sociology, some institutionalists insist that the expansion of the modern state's regulatory role imposes many practices on organizations by way of authority. Others point out that the growing professionalization of many spheres of activity is giving rise to professional communities endowed with sufficient cultural authority to impose certain norms or practices on their members[86] . In other cases, common institutional practices are supposed to emerge from a more interpretative process of discussion between the players in a given network (concerning common problems, their interpretation and resolution), and taking place in various forums, ranging from management

[84] N. FLIGSTEIN, *The Transformation of Corporate Control*, op. cit.
[85]Y. SOYSAL, *Limits of Citizenship*, op. cit.

[86] P.DIMAGGIO, W.W. POWELL, "The Iron Cage Revisited: Institutional Isomorphism and Collective Rationality" and W.W. Powell, "Expanding the Scope of Institutional Analysis, in W.W. Powell, P.J. DiMaggio, *The New Institutionalism inOrganizational Analysis*, op. cit. chapters 3 and 8.

schools to international colloquia. Such exchanges are expected to provide actors with shared cognitive schemas, which give concrete form to the intuition of appropriate institutional practices, which are then widely disseminated. In such cases, the interactive and creative dimensions of the process by which institutions are socially constituted are clearly apparent[87] . Some argue that we can even observe these processes on a transnational scale, where the usual concepts of modernity confer a certain degree of authority on the practices of the most "developed" states, and where the exchanges that take place under the aegis of international regimes encourage agreements that spread common practices beyond national borders[88] .

Within the framework of this research, sociological neo-institutionalism will enable us to understand the dynamics and instruments used by the Commune d'Arrondissement de Yaounde II to implement its local development and international cooperation policies.

VIII. METHODOLOGICAL FRAMEWORK

According to Madeleine GRAWITZ, method is "*the set of intellectual operations by which a discipline seeks to reach the truth it pursues, demonstrates it and verifies it (...). It is more or less linked to a philosophical position*"[89] . This methodological framework is therefore an

[87] On this point, we are indebted to the penetrating analysis developed by J.L. CAMPBELL in *"Recent Trends in InstitutionalAnalysis"*, p. 11.

[88] Cf. J.W. MEYER et al, "Ontology and Rationalization", J.W. MEYER, "RationalizedEnvironments", and D. STRANG, J.W. MEYER, "Institutional Conditions for Diffusion", in W.R. SCOTT, J.W. MEYER, *InstitutionalizedEnvironments and Organizations*, op. cit. chapters 1, 2 and 5.

[89] Madeleine GRAWITZ, *op. cit,* pp.351-352.

opportunity for us to present our methodological orientation, consisting of data collection methods (A), and the methodological approach adopted (B).

A. DATA COLLECTION METHODS

Our research will be based on the data collection techniques usually used in public policy analysis: documentary research (1) and interviews (2).

1. Document consultation

It is a non-reactive data collection technique, and is fundamental to our research insofar as it will enable us to gather a large amount of information.

We'll be working with documents such as strategic planning documents relating to the local development policies of the Yaounde II Commune d'Arrondissement, and the various agreements between the Commune d'Arrondissement de Yaounde II and its international and local partners.

2. The interview

Interviewing is a lively data-gathering technique and is very useful for gathering information on public policy. These interviews will enable us to get in touch with the actors involved in the processes analyzed in this study, such as managers in charge of cooperation and local development at the Commune d'Arrondissement of Yaounde II.

B. DATA ANALYSIS METHODS: THE POLICY NETWORK APPROACH

The methodological approach adopted for the analysis of data collected during our scientific investigation is the public policy network approach. After presenting it (1), we will show how it is applied to our object of study (2).

1. Method presentation

As public policy analysis has become more professionalized and freed from political control, it has developed its own controversies and questions, as well as its own analytical tools. One of the most talked-about developments has been the proliferation of players considered legitimate to participate in public action. This has given rise to issues of coordination in public action, and the conditions for participation by players from both the private and public sectors. Among the many concepts that have emerged to analyze these developments are public policy networks. The notion first appeared in Anglo-Saxon literature in the 1980s, and spread to France in the 1990s. How did the term "public action network" emerge in the public policy analyst's toolbox? Networks raise questions. They are as much a word used by players to designate their cooperation, as an analytical concept, and a method of analysis whose common denominator is a questioning of the participants in public action and the procedures they use to achieve a lasting influence on the course of public policies affecting them. The notion of "networks" has gained currency in public policy analysis, to designate the multiple actors involved in the production of public goods. It is part of a structuring debate in public policy analysis that traditionally sees neo-corporatism and pluralism in opposition. The debate on public action networks raises two major questions: Who is authorized to participate in public policy? Are the decisions taken democratic?

In the United States, theories of pluralism were dominant in the 1950s. They postulate a diversity of interests with access to the political system. Neo-corporatists sharply criticized them, and popularized the notion of the "iron triangle". To analyze industrial policies in the United States, Theodore LOWI[90] and, following him, Guy PETERS[91] describe "sub-systems" in which representatives of interest groups, state agencies and the US Congress maintain "symbiotic" relationships, defending similar interests. These exchanges escape the public eye. This notion emphasizes the closed nature of decision-making circles and the predominant role played by private players.

Discussing the notion of the "iron triangle" as too restrictive, Hugh ECLO prefers that of the "thematic network", which he defines as "*a communication network of all actors interested in political action in an area, including government authorities, legislators, businessmen, representatives of pressure groups and even academics and journalists. Such a network is obviously not an iron triangle*"[92] .
The thematic network is therefore the antithesis of the "iron triangle", and describes the circulation of ideas between different players, enabling policy changes. It brings pluralist theses back to the fore.

---

[90] Theodore LOWI, *The End of Liberalism*, New York, N. Y., Norton, 1969.
[91] Guy PETERS, *American Public Policy*, Basingstoke, Mac Millan, 1986
[92] Hugh HECLO, "Issue Network and The Executive Establishment" in Anthong King (ed.), *The New American Political System*, Washington DC, American Enterprise Institute, 1978

The public policy network approach represents a form of synthesis between these two currents. The aim was to create a middle ground between so-called pluralist theories, defending the idea of a broadly open game in which everyone can access public resources, and neo-corporatist theories, for which the game is closed, with public policies being decided within an "entre-soi" mixing public and private players. The idea is to take seriously the obvious openness of public action to multiple actors - which neo-corporatist approaches neglected, making the state a central actor capable of selecting its interlocutors - while taking into account the fact that public policies and the spaces in which they are discussed are neither public, nor accessible to all. The analysts' aim was to think about and, above all, conceptualize relations between the state, local authorities, elected representatives and interest groups in a context of profound institutional transformations (decentralization reforms, multiplication of supranational bodies, economic standards designed to reduce public spending, etc.).

The notion of a network is based on a strong observation: more or less enclosed spaces integrate public players (government departments, public establishments, agencies, etc.) and private players (companies, associations, trade unions, etc.), whose interests are manifold (for example, the quest for profitability, a public service mission, equal access to a collective good, etc.). The production of public goods is the result of exchanges between these players.

In most cases, several groups representing a variety of interests are involved in public policy. Environmental protection policies are often cited as an example of this mix: they bring together representatives of industrial groups, nature conservation associations, government departments, employees and local residents, all of whom have different reasons for taking action and different objectives. The network approach seeks to analyze the participation of these multiple groups in public policy. Widening the circle of participants is not synonymous with equality between them, but relies on asymmetries of resources between the various protagonists.

Researchers have formalized typologies, classifying public action networks according to their stability and degree of openness to new participants. Among the best-known are 'project networks', 'public policy communities' and 'epistemic communities'. As an analytical tool, the term "networks" designates intermediate spaces, places within which public action is constructed. It makes it possible to undertake a sociology of interacting public authorities, in order to move beyond analyses that placed government departments at the center, and which delivered a relatively disembodied vision of public action, reifying public power. One of the contributions of "public action networks" is to emphasize the collective, interactive dimension of decision-making processes. The result is a new way of conceiving the role of public authorities. They do not *a priori* have a central place, and may even be relegated to the status of secondary players. Networks are therefore also born of this observation of the relative disappearance of public authorities and the importance of intermediary spaces in the structuring of public action.

2. Its application to the object of study

The network approach to public policy will thus enable us to analyze the various interactions between the different actors involved in the sphere of local development and cooperation in the Yaounde II district: the State, whose action is materialized by the Ministry in charge of Decentralization and the other partner ministries of the Yaounde II District Commune, its international partners, and local people.

PART ONE: SOCIOGENESIS, NORMATIVE AND INSTITUTIONAL DYNAMICS OF DECENTRALIZATION IN CAMEROON

Over the past two decades, decentralization has become a political priority for many African states. Against a global backdrop of revaluation of the local, redefinition of the State, economic and financial crisis and pressure from donors, central African governments have embraced this new way of organizing public action. Cameroon, for example, has adopted decentralization as a fundamental mode of state management. The constitutional revision of January 18, 1996 proclaims that the Republic of Cameroon is a "decentralized unitary state". Added to this are the laws of July 22, 2004 on the orientation of decentralization, which grant greater prerogatives to local authorities than in the past. This highlights the importance of communes and regions as territorial echelons where affairs should be self-managed. This raises questions about the mechanisms for legal and institutional appropriation of this mode of administration by the State of Cameroon.

To this end, the first part of this study focuses on the sociogenesis and institutional and normative dynamics of decentralization in Cameroon. With this in mind, it is divided into two chapters, the first of which focuses on the historical context and legal framework of decentralization in Cameroon (Chapter I), while the second highlights the main actors and institutions supporting decentralization in Cameroon (Chapter II).

CHAPTER I: HISTORICAL BACKGROUND AND LEGAL FRAMEWORK OF DECENTRALIZATION IN CAMEROON

In the legal theory of the State, where the concept of decentralization finds its best application, it refers to the recognition, alongside the State, of public bodies entrusted with administrative responsibilities[93] . These entities enjoy relative autonomy in decision-making and management, but act under the supervision of the state[94] . Basically, then, it's a question of the degree of autonomy granted by the state to sub-state bodies. Despite the complexity of socio-historical considerations, the origins of decentralization, and more specifically of the communal movement in Cameroon, can be traced back to 1916, a period marked by the German colonial administration[95] . In this chapter, we will first highlight the chronological evolution of the decentralization process in Cameroon since its origins (Section I), then review the literature on the legal and normative framework of decentralization in Cameroon (Section II).

SECTION I: CHRONOLOGICAL DEVELOPMENT OF DECENTRALIZATION IN CAMEROON

The chronological evolution of the decentralization process in Cameroon can be divided into two main phases: an initial or even pre-initial phase (Paragraph I), and an appropriation and operationalization phase (Paragraph II).

---

[93] Landry NGONO TSIMI - Cours de *Collectivités locales, décentralisation et droit de la coopérationdécentralisée au Cameroun*, Master "Coopération internationale, Action humanitaire et Développementdurable", Yaoundé, Institut des Relations internationales du Cameroun: 2015, p. 5

[94] Michel VERPEAUX, *Droit des collectivités territoriales*, PUF, 2005, pp. IX et seq.

[95] Landry NGONO TSIMI, *L'autonomie administrative et financière des collectivités territoriales et décentralisées : l'exemple du Cameroun*, Doctoral thesis defended at the Université de Paris-Est, 2010, p. 32

PARAGRAPH I: THE PRE-INITIAL (1920-1974) AND INITIAL (1974-1996) PHASES OF DECENTRALIZATION IN CAMEROON.

In this period, the pre-initial phase runs from 1916 to 1974, and the initial phase from 1974 to 1996.

A- THE PRE-INITIAL PHASE: FROM 1916 TO 1974.

During this phase, Germany, the great loser of the First World War, was forced to abandon its colonial possessions, which would henceforth be administered by the League of Nations, in accordance with the Treaty of Versailles of June 28, 1919. This gave the British and French powers a mandate to administer Cameroon. The communal movement was thus marked by the influence of these two powers. Indeed, while the British power implemented the indirect rule system, the French colonial administration implemented direct administration[96] . Thus, in 1944, there were more than thirty communes in West Cameroon, a number that rose to 28 in 1967, then 24 again in 1967, then 24 in 1969, a number that remained unchanged until the adoption of the 1972 Constitution and the law on communal organization of the United Republic of Cameroon of December 5, 1974[97] . In East Cameroon, the first communes were created by the French Governor of Cameroon's decree of June 25, 1941, which created the mixed communes of Douala and Yaoundé[98] in the country's two largest conurbations.

[96]Landry Ngono TSIMI, *op. cit.*

[97] Landry NGONO TSIMI, *L'autonomie administrative et financière des collectivités territoriales et décentralisées : l'exemple du Cameroun*, Doctoral thesis defended at the Université de Paris-Est, 2010, p. 36

[98]*Ibid.*

After the Second World War, a decree of August 21, 1952 created mixed rural communes and extended them to all subdivisions of the country. Three years later, French law no. 55-1489 of November 18, 1955 on municipal reorganization in Black African countries, with the exception of Senegal, introduced full-function communes and medium-function communes[99] . In 1967, the law of March 1 created the first "special-regime" variants within the full-function communes, in the major cities of Douala, Yaoundé and Nkongsamba. These communes would later become urban communities, headed by government delegates appointed by the President of the Republic[100] . At the dawn of Reunification, Cameroon had 339 communes under the aegis of the Communal Law of December 5, 1974[101] .

B- THE INITIAL PHASE FROM 1974 TO 1996

The unification of Cameroon in 1972 was accompanied by a series of reforms, the most important of which involved the multiplication and standardization of administrative districts by the law of July 24, 1972[102] . At local level, this standardization led to a strengthening of central authority and the replacement of regions, subdivisions and administrative posts by provinces, departments, arrondissements and districts respectively. The administrative division of the territory, and in particular the arrondissements, served as the basis for the creation of communes; Cameroon then had 339 communes under the aegis of the Communal Law

[99] *Ibid.*

[100] Landry Ngono TSIMI, *op. cit.*

[101] *Ibid.*

[102] Landry NGONO TSIMI - Cours de *Collectivités locales, décentralisation et droit de la coopérationdécentralisée au Cameroun*, Master "Coopération internationale, Action humanitaire et Développementdurable", Yaoundé, Institut des Relations internationales du Cameroun: 2015, p. 17

ofDecember 5, 1974 and its various amending texts[103] . The Constitutional Act of January 18, 1996 created a new tier of territorial authorities (the Region) on the basis of the ten existing provinces.

Until the promulgation of the decentralization laws of July 22, 2004, the law of December 5, 1974 remained the communal charter in force in Cameroon. This law reflected the Constituent's spirit of resistance to anything that might hinder the smooth functioning of national unity. Even its successive amendments have not improved the autonomy of decentralized local authorities. On the contrary, by reinforcing the powers of guardianship and appointment of local executives, it marked a clear step backwards in relation to the 1955 Communal Law, which instituted the election of mayors by municipal councils in the largest communes[104] . While the 1972 Constitution placed the organization of local authorities within the domain of the law (article 20, paragraph 3), thus implicitly recognizing their existence, the law of December 5, 1974, in turn, delegated these same powers to the executive. Through this twofold transfer of powers, and under the leadership of the single party, central government had finally created and organized the commune as a decentralized administrative unit. Local executives were appointed and local assemblies co-opted from the single party. The powers devolved to the communes were tightly controlled under the local impulse of the Party-State association. It was therefore difficult to evoke the very idea of local government autonomy, in the sense given to it by modern doctrine[105] . Since then, Cameroon has seen the advent of a multi-party system and pluralist elections. At local level, the law of August 14, 1992 replaced all

---

[103]*Ibid.*

[104]*Ibid*, p. 20

[105]Landry NGONO TSIMI, *op. cit.* p. 21

"municipal administrators" in rural communes, who were appointed by the mayors elected by their municipal councils, even though this law did not come into force until the municipal elections of January 21, 1996, four years later.
The new Constitution of 1996 marked a break with single-party rule and the slogan of "national unity"[106] . Numerous fundamental freedoms have since been recognized by the legislator.

PARAGRAPH II: THE OWNERSHIP (1996 TO 2004) AND LEGAL OPERATIONALIZATION (2004 TO PRESENT) PHASES

In this section, we will illustrate the different dynamics that have led to the appropriation and legal operationalization of the concept of decentralization by the State of Cameroon.

A- THE LEGAL APPROPRIATION PHASE: 1996 TO 2004

The Constitutional Law of January 18, 1996 created a new tier of territorial authorities (the Region) on the basis of the ten existing provinces. To date, there are 384 decentralized local authorities in Cameroon, as a result of the decentralization laws of July 22, 2004 and the numerous arrondissements created in the meantime[107] . July 22, 2004: the first, n° 2004/017, concerns the orientation of decentralization; the second, n° 2004/018, lays down the rules applicable to communes, and the third, n° 2004/018, lays down the rules applicable to regions. Unlike previous legislation (Communal Act of Dec. 5, 1974), article 152 of Act no. 2004/017 institutes a single nomenclature for the commune, thus doing

[106]*Ibid.*
[107]see decree 12/11/2008

away with the distinction between urban and rural communes. Given that many agglomerations were made up of urban and rural zones, and therefore of urban and rural communes in the same town (for example, the urban commune of Kribi, and the rural commune of Kribi, in the town of the same name), the implementation of this provision entailed either the grouping together of certain communes within an urban community, or the geographical relocation of the chief town of one of the two communes. This, moreover, seems to be the meaning of article 153 of the Communal Law: "Communes having their chief-place on the territory of another commune have 18 months ... to transfer the said chief-place to their territory"[108] .

B- THE LEGAL OPERATIONALIZATION PHASE: AFTER 2004

At the time of the 1996 municipal elections, Cameroon had a total of 338 communes, including 2 urban communities, 9 communes under a special regime, 11 urban arrondissement communes and 306 rural communes. It was not until 2004 that a law on decentralization was passed, giving new impetus to decentralization in Cameroon. Since 2008, the number of communes has risen from 339 to around 360, with 10 urban communities in each regional capital.

SECTION II: NORMATIVE AND REGULATORY FRAMEWORK FOR DECENTRALIZATION IN CAMEROON

[108]Landry NGONO TSIMI, *op. cit.* p. 17

This section highlights the textual dynamics (Paragraph I) and fundamental principles (Paragraph II) of the legal framework for decentralization in Cameroon.

PARAGRAPH I: CHRONOLOGY OF DECENTRALIZATION LEGISLATION IN CAMEROON

In this paragraph, we will attempt to show how the textual dynamics relating to the legal framework for decentralization in Cameroon obey two main phases. Firstly, a conceptualization phase marked by a proliferation of legal texts, and secondly, a phase marked by a political will to strengthen this framework.

A - A PHASE OF CONCEPTUALIZATION OF THE LEGAL FRAMEWORK MARKED BY A PROLIFERATION OF TEXTS: FROM 2004 TO 2011

Cameroon's decentralization policy is enshrined in the Constitution and supported by a coherent legal arsenal:

- Law n° 92-002 of August 14, 1992 setting the conditions for the election of municipal councillors. Amended and completed by law n° 2006/010 of December 29, 2006;

The law of January 18, 1996, revising the Constitution of June 2, 1972, established the Republic of Cameroon as a decentralized unitary state. This gave decisive impetus to the decentralization process in our country;

- Law n° 2004/017 of July 22, 2004 on the orientation of decentralization ;

- Law no. 2004/018 of July 22, 2004 establishing the rules applicable to communes ;

- Law no. 2004/019 of July 22, 2004 establishing the rules applicable to the regions ;
- Law n° 2006/005 of July 14, 2006 setting the conditions for the election of senators ;
- Law n° 2006/004 of July 14, 2006 establishing the mode of election of regional councillors;
- Law n° 2006/011 of December 29, 2006 on the creation, organization and operation of "Elections Cameroon" (ELECAM);
- Loi N°2009/11 Du 10 Juillet 2009 Portant Régime Financier Des Collectivités Décentralisées ;
- Law n°2009/019 of December 15, 2009 on local taxation.
- Décret No 2010/1735/PM du 01 juin 2010 fixant la nomenclature budgétaire des collectivités territoriales décentralisées.

In addition to this raft of laws, the Ministry of Territorial Administration and Decentralization also drew up a draft statute for communal personnel: Arrêté n°00136/A/MINATD/DCTD of August 24, 2009, bringing into force the standard tables for communal jobs. As well as the definition of the government's urban strategy, which aims to highlight and strengthen the role of decentralized local authorities in urban management, mainly in the fields of land and urban planning.

B - A PHASE MARKED BY THE POLITICAL WILL TO STRENGTHEN AND COMPLETE THE LEGAL FRAMEWORK: THE GENERAL CTD CODE OF 24/12/2019.

The law on the General Code of Decentralized Territorial Authorities (Code Général des Collectivités Territoriales Décentralisées - CTD) was adopted by Parliament and promulgated by the President of the Republic on December 24, 2019. It contains major innovations. The Code enshrines :

- a little less omnipresent guardianship
- more competences transferred to CTDs than in the past
- greater functional autonomy
- more resources transferred directly to CTDs
- promoting the participation of local people in drawing up the budget and selecting priority projects, through neighbourhood and village representatives
- an advanced definition of the status of local elected representatives
- greater clarification of responsibilities between the Urban Community and the district commune
- elections to the executive bodies of urban communities
- a more precise financial regime

The Code also enshrines a Special Statute for the North-West and South-West regions to boost development in these two regions, with a view to improving the well-being of their populations. It takes into account linguistic heritage and *common law*, and provides for :

- a special financial endowment
- advantageous tax treatment
- special bodies to manage the Region
- A mediator to solve certain problems or settle certain issues.

Last but not least**, the** disappearance of the position of Government Delegate has been formalized in the Code. From now on, there will be an elected mayor of the city, known as the super-mayor. He or she will be a native of the region to which the city belongs. He is elected. He must be a municipal councillor to be eligible**.**

In accordance with the provisions of the aforementioned Code, the President of the Region is also an indigenous person who is a national of the Region. Similarly, the Vice-Chairman of the Region is an indigenous person and a traditional chief. It should also be noted that the Mayor is a resident of the district, and the candidates on the lists must reflect the sociological composition of all these elections.

PARAGRAPH II: LEGAL PRINCIPLES RELATING TO THE IMPLEMENTATION OF DECENTRALIZATION IN CAMEROON

These are the principles governing the implementation of decentralization in Cameroon. These are the principles of subsidiarity (A), equality and progressiveness (B).

A - THE PRINCIPLE OF SUBSIDIARITY

Given that most local authorities have limited resources, this principle must be applied pragmatically.

B - THE PRINCIPLES OF EQUALITY AND PROGRESSIVENESS

The principle of equality refers to the State's ability to reduce all local authorities to the same denominator. To make decentralization more operational throughout the country, the State transfers the same powers to all local authorities in the same category.

As for the principle of progressivity, it postulates that the distribution of powers should take into account the capacity of local authorities to exercise them.

In conclusion, the evolution of the decentralization process in Cameroon follows a chronological logic marked by major phases. Indeed, if the first phases that saw the advent of this process can be qualified as preliminary, namely a pre-initial phase running from 1916 to 1974, and an initial phase running from 1974 to 1996, as well as a phase of legal appropriation running from 1974 to 1996 and a phase of operationalization, the evolution of the normative framework has also undergone two major phases, namely a progressive implementation of texts, and then a political will to reinforce this legal framework. Thus, while based on the principles of subsidiarity, equality and progressiveness, decentralization in Cameroon appears to be an ongoing process. To better understand the construction of this process, however, we need to examine the actors involved and their various roles.

CHAPTER II: ACTORS AND INSTITUTIONS SUPPORTING DECENTRALIZATION IN CAMEROON

The new direction given to the decentralization process has led to a restructuring of the country's institutional architecture. The State has set up structures to monitor the development and operation of decentralized local authorities. Bodies were created and institutions reorganized to meet the new demands of decentralization. Today, decentralization actors can be found at both State and societal levels. Thus, among the actors involved in the decentralization process in Cameroon, a distinction can first be made between central and deconcentrated services on the one hand (Section I), and monitoring and support institutions on the other[109] (Section II).

SECTION I: CENTRAL AND DECENTRALIZED SERVICES

The aim here is to highlight the different roles of central and decentralized services in the decentralization process in Cameroon.

PARAGRAPH I: THE KEY PLAYERS IN DECENTRALIZATION IN CAMEROON

By illustrating the dynamics of the central players in decentralization, we can see the involvement of the President of the Republic and the Prime Minister, the ministry in charge of decentralization and other ministerial departments.

---

109 Landry NGONO TSIMI - Course on *local authorities, decentralization and cooperation law in Cameroon*, Master's degree in "International Cooperation, Humanitarian Action and Development durable", Yaoundé, Institut des Relations internationales du Cameroun: 2015,p.4

The Constitution clearly states that the President of the Republic defines national policy[110] . In accordance with the laws of decentralization, the powers of the State over decentralized local authorities are exercised by the President of the Republic. As such, he can decide on the temporary grouping of certain communes, or declare the attachment of one commune to another, or even its break-up. As part of the exercise of his prerogatives, on April 24 and 25 2007, he signed two decrees respectively creating fifty-eight (58) new communes and setting the number of municipal councillors per commune. Similarly, in January 2008, he issued decrees creating twelve (12) new urban communities, headed by Government Delegates appointed in early 2009.

As part of his supervisory role, he also has the power to dissolve the municipal council and dismiss the mayor and his deputies in the event of gross misconduct or violation of current laws and regulations. This was the case for the Mayor of the commune of Penja, who was dismissed by decree no. 2008/193 of June 2, 2008, concerning the dismissal of a municipal magistrate.

As Head of Government, the Prime Minister plays a decisive role in implementing the decentralization process, based on his general mission of coordinating government action[111] .

For example :

- In 2008, a PM circular dated January 11 instructed the heads of ministerial departments to take decentralization into account in their respective sectoral strategies.

---

[110]Republic of Cameroon, *Constitution of January 18, 1996*, Title II, Art 5, p. 5

[111]Landry NGONO TSIMI, *op cit*, p. 5

- in 2008, a PM decree specified certain organizational and operational procedures for the deliberative bodies and executives of communes, urban communities and local authority associations.
- He is an essential link in the implementation of the decentralization process. He chairs the National Decentralization Council, whose Permanent Secretariat is provided by a senior member of his staff.

B - THE MINISTER FOR LOCAL AUTHORITIES AND OTHER MINISTERIAL DEPARTMENTS

The laws of 2004 gave the Minister responsible for local authorities a decisive role in leading and monitoring the decentralization process[112] , and these prerogatives were reiterated in the new general code for decentralized local authorities[113] .

In the exercise of his supervisory powers, the mayor may, by reasoned decree, suspend a municipal councillor, the mayor or deputy mayors, or the entire municipal council, in the cases provided for by the law laying down the rules applicable to municipalities.

If a municipality is unable to function, it can set up a special delegation to carry out the functions of the municipal council. This was the case in the municipality of Lobo, following the outcome of the pre-electoral dispute that began in July 2007.
The Minister in charge of decentralized local authorities also plays a decisive role in the operation of municipal and urban community bodies.

[112]Landry NGONO TSIMI, *Ibid*, p. 5
[113] Republic of Cameroon, *CTD General Code*, December 24, 2019.

It is the Minister who sets, by decree, the amount of the sessional allowance and the reimbursement of expenses allocated to the mayor, deputy mayor, municipal councillor, president and member of the special delegation, in the performance of their duties. Likewise, it approves the municipal council's decision to set the remuneration and functional and representation allowances of the mayor and deputy mayors.

Pursuant to the provisions of article 75 paragraph 1 of law no. 2004/018 of July 22, 2004 laying down the rules applicable to communes, the Minister in charge of local authorities made the standard tables of communal jobs enforceable by order dated August 24, 2009, with a view to boosting the efficiency and effectiveness of services provided by communes and urban communities, while taking into account the importance of each in the development process at local level.

He also appoints Town Hall Secretaries-General and terminates their functions. He also appoints Municipal Receivers by joint decree with the Minister of Finance.

It also authorizes the recruitment of municipal staff from the seventh category upwards, and approves their employment contracts. It also approves contracts for the concession of municipal industrial and commercial public services, cooperation agreements with foreign local authorities, and membership of international organizations of twinned towns or other international organizations of towns[114] .

As part of the effective implementation and monitoring of decentralization, the Minister responsible for local authorities chairs the Comité Interministériel des Services Locaux. It is in this capacity that this committee has assisted the various ministerial departments concerned by

[114]Landry NGONO TSIMI, *Ibid,* p.5

the transfer of powers in determining the first powers to be transferred[115]
.

The range of powers transferred to communes and urban communities by the 2004 decentralization laws calls for the involvement of other ministerial departments, which are also members of the Comité Interministériel des Services Locaux and the Conseil National de la Décentralisation. In this context, the Ministry of Finance and the Ministry of Public Investment can be singled out for their general role in financing decentralization and monitoring communal projects.

These ministerial departments, whose actions complement those of the Ministry in charge of decentralization, were logically joined by the nine (09) that have effectively transferred competencies and resources to communes and urban communities since fiscal 2010, the year that marks the starting point of the first effective transfers. These are[116] :

- Ministry of Social Affairs ;
- Ministry of Agriculture and Rural Development ;
- Ministry of Culture ;
- Ministry of Basic Education ;
- Ministry of Livestock, Fisheries and Animal Industries ;
- Ministry of Energy and Water ;
- Ministry for the Promotion of Women and the Family ;
- Ministry of Public Health ;
- Ministry of Public Works.

[115]Landry NGONO TSIMI, *Ibid,* p 6.
[116]*Ibid*, p. 7

In addition to all the ministerial departments mentioned above, the role of the Ministry of External Relations in implementing and monitoring international decentralized cooperation needs no introduction.

PARAGRAPH II: DECENTRALIZED AUTHORITIES AND LOCAL PLAYERS

Indeed, we can't talk about decentralization without talking about grassroots players. These are the players involved in the process of implementing local development as close to the people as possible. They include the authorities representing the central state, i.e. the decentralized authorities, and the local authorities, who are responsible for managing the local population.

A- DECENTRALIZED AUTHORITIES

Decentralization, as a way of organizing the unitary State, cannot be separated from deconcentration, in which decentralized authorities act as local relays for central authorities. Indeed, decentralization laws specify their role in the implementation of the decentralization process, notably through advisory support to decentralized local authorities.

In accordance with the decentralization laws of July 22, 2004[117] , the Governor is the State's guardian of the regions. This is what justified and explains, if need be, the creation of a Regional Development Division as part of the restructuring of the regional governor's services under the

[117]Landry NGONO TSIMI, *Ibid,* p. 8

aforementioned decree of November 12, 2008. Until 2010, he still supervised the communes through the former provincial communes departments, notably as regards the management of communal personnel, oversight of the operation of municipal councils, and approval of budgets and administrative accounts[118] .

The Préfet, representing the State in the département, is responsible for overseeing the communes and their establishments. He alone is empowered to speak on behalf of the State before the municipal councils of the communes within his jurisdiction. The decentralization laws of 2004 gave this administrative authority significant powers to guarantee and ensure the smooth running of communes. With this in mind, as part of the 2008 reform of the organization and operation of the Prefecture's services, a Local Development Service was created, tasked with assisting the Prefect in exercising the State's supervisory authority over communes and communal public establishments. The creation of this department is of the utmost importance, given its mission to promote and monitor development initiatives at local level. In fact, this service is called upon to make the performance of communes and their establishments more legible through its advisory support for their harmonious operation, budgetary control and legality control of their acts, as well as the monitoring of decentralized cooperation.

In view of the role they are called upon to play in the implementation of decentralization, the Prefects have taken part in numerous seminars and workshops organized for them or for municipal magistrates, with a view to strengthening their capacity to support communes. The Prefect approves most municipal acts, as well as initial and supplementary

[118]Landry NGONO TSIMI, *Ibid*, p. 8

budgets, off-budget accounts and special expenditure authorizations. Prior to their adoption, the Prefect approves communal development plans, and can annul manifestly illegal communal acts, particularly in cases of right-of-way or voie fait. Likewise, it may refer to the competent administrative court any acts of the mayor or municipal council that it considers to be illegal, as soon as they are received.

In addition to his role of general supervision, coordination and coordination of decentralized State services in the département, the Prefect also plays a decisive role in the conclusion of agreements and conventions drawn up on behalf of the State with the communes. He ensures that the laws are properly applied by the communes, and that they make good use of the decentralized government services[119] . In addition to the reform introduced on November 12, 2008, the regulatory framework for its intervention has been further developed through decrees relating to transfers and orders laying down specifications, particularly with regard to the monitoring and evaluation of the effective exercise of powers by communes.

Under the terms of the decentralization laws of July 2004, the Sous-préfet is not a supervisory authority of the State over the communes. However, as the representative of the State at the arrondissement level, whose territorial boundaries the commune follows, this administrative authority, hierarchically placed under the Prefect to whom it reports on its actions, is closer to the commune. This proximity enables the State to better coordinate the actions carried out by the communes as part of their mission to improve the standard of living and living environment of the population. The support provided by this administrative authority has

[119]Landry NGONO TSIMI, *op. cit,* p. 8

always proved indispensable in many respects. This is particularly true when it comes to collecting taxes, and exercising municipal policing powers. This support is even more justified in view of its responsibilities for economic development in its administrative district.

As part of the exercise of the powers transferred by the State to the communes, the State's decentralized services must provide them with the technical support and assistance needed to build the works and carry out the related activities, while ensuring compliance with the standards in force.

The decrees setting out the terms and conditions for exercising these powers, and the specifications specifying the technical terms and conditions, stipulate that these services must draw up a half-yearly report on the implementation of the transferred powers.

This support and technical assistance is all the more necessary given that some of the powers transferred to communes and urban communities are highly technical, requiring qualified human resources that the local authorities concerned do not always have. Only the decentralized departments of the State have such resources for certain matters.

B - LOCAL PLAYERS

Decentralized local authorities are structured around two key bodies: the executive and the deliberative[120] bodies. Legally, these bodies are elected by direct universal suffrage for municipal councillors, who in turn elect mayors and their deputies from among their number.

---

[120] République du Cameroun, *Code Général des CTD*, Titre III, p. 42 - 47.

All municipal councillors form the municipal council, which is the deliberative body of the commune. To ensure the smooth running of the council, the creation of committees is governed by the provisions of decree no. 2008/0752/PM of April 24, 2008, specifying certain organization and operating procedures for the deliberative bodies and executives of the commune, the urban community and the syndicate of communes.

Municipal magistrates, i.e. district mayors and city mayors, represent the commune or urban community, as the case may be, in civil and legal matters.

At commune level, the mayor and deputy mayors constitute the commune's executive body. They are elected by the municipal councillors for the same term as the municipal council[121] . The mayor is elected by a two-round uninominal majority ballot. Once the mayor has been elected, the deputy mayors are elected by proportional representation on the basis of the highest average vote[122] .

It should be noted that the election of mayors and deputy mayors may be the subject of annulment proceedings, in accordance with the rules laid down by current legislation for the annulment of the election of municipal councillors. For example, the 2007 municipal elections gave the Administrative Chamber of the Supreme Court the opportunity to rule on the election of the mayors of Biyouha and Mbanga[123] . The latter case was definitively annulled, resulting in the organization of a new election.

[121]*Ibid.*

[122]*Ibid.*

[123]Landry NGONO TSIMI, *op. cit,* p. 10

Added to this landscape are the regional councillors, whose elections were held for the very first time in Cameroon, on December 06, 2020. In addition, decree no. 2021/043 of January 25, 2021 appointed the secretaries-general of the regional councils.

SECTION II: BODIES RESPONSIBLE FOR MONITORING, SUPPORTING AND CONTROLLING DECENTRALIZATION

Completing the decentralization process in Cameroon also requires the involvement of other institutions no less important than those mentioned above. The aim here is therefore to highlight, firstly, the bodies responsible for monitoring decentralization (Paragraph I), and secondly, those responsible for support and control (Paragraph II).

PARAGRAPH I: MONITORING BODIES

Title V, articles 78 and 79 of the Decentralization Orientation Act creates two (2) monitoring bodies: the National Decentralization Council and the Interministerial Committee for Local Services.

A- THE NATIONAL DECENTRALIZATION COUNCIL

The National Decentralization Council is the body responsible for monitoring and evaluating the implementation of decentralization. Its organization and operation are set out in decree no. 2008/013 of January 17, 2008 issued by the President of the Republic.

Chaired by the Prime Minister, Head of Government, this body is made up of twenty-two (22) statutory members, including members of the Government concerned by the transfer of powers, two (02) senators, two (02) deputies and two (02) representatives of the Economic and Social Council. The Chairman of the Council may invite any other person to attend, in view of his or her expertise on the issues on the agenda.

For some time now, the CND has also been open to 2 representatives. The project now opens the composition of the Council to representatives of communes, urban communities and regions.
The Council submits an annual report to the President of the Republic on the state of decentralization and the operation of local services[124] . It also issues opinions and recommendations on the annual program for the transfer of powers and resources to decentralized local authorities, and on the terms and conditions of such transfers[125] .
To carry out its missions, the Conseil National de la Décentralisation (National Decentralization Council) has a Permanent Secretariat, whose composition and organizational and operational procedures are set by order of the Prime Minister. To this end, on February 12, 2008, the Head of Government signed Order no. 022/CAB/PM establishing the composition and specifying the organization and operating procedures of the said secretariat. The Permanent Secretariat, which is coordinated by a Permanent Secretary, holds regular sessions at which issues relating to the preparation of Council sessions are discussed.
Council. Its composition provides an overview of the evolution of the decentralization process.
Since 2008, when it was set up, it has held a number of meetings, some of which have been extended to other ministerial departments that are not statutory members, to development partners and to the French and German cooperation agencies. It has also organized a number of seminars and workshops for its members, with a view to building their capacity to implement decentralization and monitor the transfer of skills and resources.

[124]Landry NGONO TSIMI, *op. cit,* p. 10.
[125]*Ibid.*

B- THE INTERMINISTERIAL COMMITTEE FOR LOCAL SERVICES

The Comité Interministériel des Services Locaux is an interministerial consultative body reporting to the Minister responsible for decentralization. Its mission is to prepare and monitor transfers of powers and resources to decentralized local authorities, as decided by the ministerial departments concerned[126] .

In line with the provisions of decree no. 2008-014 of January 17, 2008, which sets out the organization and functioning of the body, it has held several meetings since June 2008, the date of its first session.

To carry out its tasks, the Interministerial Committee on Local Services has a Permanent Technical Secretariat coordinated by the Director in charge of decentralized local authorities[127] .

The Permanent Technical Secretariat is responsible for

- receiving, recording and distributing committee mail;
- dispatch of correspondence from the committee;
- the secretariat for committee meetings;
- the preparation of files to be submitted to the Committee and the National Decentralization Council for examination;
- monitoring and evaluating implementation of the committee's guidelines and recommendations;
- preparing the committee's activity reports and action plans;
- preservation of the committee's documents and archives;
- carrying out any other tasks assigned to it by the Committee.

PARAGRAPH II: SUPPORT AND CONTROL BODIES

[126] *Ibid.*

[127] Landry NGONO TSIMI, *op. cit*, p. 11.

The delicate nature of the decentralization process has obliged public authorities to set up specific institutions whose general mission is to support the players involved in this process. What's more, local authorities cannot escape traditional control techniques and bodies.

A- SUPPORT INSTITUTIONS

Support for decentralization players is a necessity that the public authorities have been able to make a reality. Indeed, the institutional landscape of decentralization in Cameroon includes numerous bodies that support decentralization actors in one way or another. The three main ones are the Fonds Spécial d'Equipement et d'Intervention Intercommunale (FEICOM), the former Centre de Formation pour l'AdministrationMunicipale CEFAM (now the Ecole Nationale d'Administration Locale - NASLA) and the Programme National de Développement Participatif (PNDP)[128] .

FEICOM is a Public Administrative Establishment with legal personality and financial autonomy. It was created by law no. 74/23 of December 5, 1974, on the organization of communes, in a dual political and economic context marked by the constitutional reform of 1972 and the global economic crisis triggered by the oil crisis of 1973, which led to a decline in State revenues and, incidentally, those of communes. Its organization was subsequently defined by a decree dated March 25, 1977. Since its creation, FEICOM's strategy has been based on three main axes:[129] :

[128]Landry NGONO TSIMI, *op. cit,* p. 11.
[129]Landry NGONO TSIMI, *Ibid,* p. 12.

- the pooling of human resources by making qualified human resources available to all communes, with a view to supporting local elected representatives in setting up and carrying out projects;
- the pooling of technical resources by making civil engineering equipment available to municipalities grouped together in a syndicate;
- the pooling of financial resources through the management of mutual funds to finance communal and inter-communal projects.
- FEICOM's three missions at the time of its creation illustrate this vision of the public authorities. They cover :
- mutual assistance between communes through solidarity contributions and cash advances;
- financing communal or inter-communal investment projects;
- cover training costs for municipal and registry staff. Since 1998, a fourth mission has been added, namely "the centralization and redistribution of additional municipal levies". To meet the new challenges posed by the acceleration of the decentralization process, FEICOM embarked on a restructuring process in 2005, which culminated in the signature by the President of the Republic of a decree dated May 31, 2006, reorganizing the public establishment. In addition, under the terms of decree no. 2009/248 of August 05, 2009 setting out the terms and conditions for the distribution of the General Decentralization Allocation, FEICOM is responsible for making available to the beneficiary communes, syndicats de communes and urban communities, the corresponding shares of the said Allocation instituted by article 23 of the Decentralization Orientation Act of July 22, 2004. Likewise, the Local Taxation Act of December 15, 2009 entrusts it with the centralization and repayment to the Communes and Urban Communities of taxes and levies subject to equalization.

Located in Buea in the South-West region, CEFAM is a training institution for communal staff and local elected representatives created by the State in 1977. In 2020, it was nationalized and became the Ecole Nationale d'Administration Locale (NASLA) by decree no. 2020/111 of March 2, 2020. This public institution with legal personality and financial autonomy, placed under the authority of the Minister in charge of decentralization, is an instrument set up to train, upgrade and retrain the administrative and technical staff of communes, commune syndicates and communal establishments, as well as the staff responsible for overseeing communes and the staff in charge of civil status. CEFAM provides training in three (3) cycles:

- Cycle I, designed to train municipal administration executives;
- Cycle II, designed to train local authority staff;
- Cycle III, designed to upgrade and retrain municipal staff.

To this end, the company has taken part alongside other government bodies in the organization of numerous symposia, seminars and workshops to build the capacity of magistrates and municipal staff, as well as other players in the decentralization process.

Created in 2004, the PNDP is a multi-donor program designed to assist the Cameroonian government in promoting growth and job creation for the sustainable development of rural communities[130] . It aims to define and implement mechanisms to empower communes and their communities at grassroots level, making them actors in their own development, as part of the progressive decentralization process[131] . The PNDP was designed in three phases, each lasting four years. The first phase, launched in 2004

[130] www.pndp.org visited on January 24, 2021 between 5:00 pm and 6:00 pm.
[131] *Ibid.*

and completed in 2009, covered 155 communes in the six regions of Adamaoua, Centre, Extreme-Nord, Nord-Ouest and Sud. The second phase covered all 10 regions and 329 communes, and was officially launched on January 29, 2010. The 3ème phase covers all communes, including arrondissement communes.

B - CONTROL BODIES

Numerous institutions are involved in the effective and efficient implementation of decentralization and, more specifically, the exercise of transferred powers, through technical controls. Such is the case of the Contrôle Supérieur de l'Etat and the technical structures of certain ministerial departments[132] .

Superior State Audit In accordance with the provisions of decree no. 2005/374 of October 11, 2005, which organizes the departments of the

They report directly to the President of the Republic, from whom they receive instructions and to whom they are accountable.

Cameroon's supreme audit institution is headed by a Minister Delegate of the Presidency of the Republic. In charge of external audit. Its missions include

- high-level auditing of public services, public establishments, decentralized local authorities and their establishments, public and semi-public companies, control of state budget execution;
- monitoring the execution of externally-funded projects;
- project and program evaluation ;
- technical, methodological and educational support for controlling and auditing the management of public assets;

[132]Landry NGONO TSIMI, *op. cit*, p. 13.

Minister Delegate CONSUPE presides over the Conseil de Discipline Budgétaire et Financière, a supreme audit institution. Under the terms of decree no. 2008/028 of January 17, 2008, which organizes its operation, this non-jurisdictional body sanctions irregularities and mismanagement committed by Mayors and Government Delegates in the performance of their duties.

Checks carried out by the Minister responsible for local authorities

The organization chart of the Ministry in charge of CTDs has given this ministerial department two (02) technical structures specifically in charge of CTD control. These are the Decentralized General Inspectorate and the Control Brigade[133] .

Jurisdictional control of decentralized local authorities is essentially exercised by the courts, in connection with the establishment or operation of municipal bodies. Because decentralization is enshrined in the fundamental law, these courts are empowered by the Constitution, the laws on judicial organization, and the General Code of CTDs, to hear cases concerning acts carried out by communes. They are[134] :

- administrative jurisdictions ;
- of the courts of audit.

In addition to the above-mentioned institutions, there are the administrative courts, comprising the administrative tribunals and the Administrative Chamber, which exercise varying degrees of control over

[133] At the time of the former Ministry of Territorial Administration and Decentralization, this control was carried out by the General Inspectorate in charge of CTDs, but with the creation of the Ministry in charge of Decentralization and Local Development, control is now carried out specifically by the respective departments in line with their competencies.

[134]Landry NGONO TSIMI, *op. cit,* p. 13.

the establishment and operation of municipal bodies, in the context of litigation management[135] .

Similarly, the local taxation law of December 15, 2009 entrusts it with the centralization and repayment to Communes and Urban Communities of taxes and levies subject to equalization. Litigation concerning the election of municipal councillors falls within the jurisdiction of the administrative courts in the first instance, and of the Administrative Chamber of the Supreme Court on appeal. The same administrative court is competent to hear disputes arising from the election of the mayor and his deputies, in application of law no. 92/003 of August 14, 1992 setting the conditions for the election of municipal councillors, amended and supplemented by law no. 2006/010 of December 29, 2006. However, pending the establishment of administrative courts, the Administrative Chamber of the Supreme Court is currently responsible for administrative litigation.

Following the example of the administrative order, the courts of audit include, on the one hand, the Audit Chamber of the Supreme Court, which is the apex body in this field, and on the other hand, the lower courts of audit[136] .

Under the terms of law no. 2003/005 of April 21, 2003 setting out the remit, organization and operation of the Audit Chamber of the Supreme Court, the latter is empowered to audit and rule on "the accounts or documents in lieu thereof of public accountants or de facto public accountants":

- the State and its public establishments ;
- decentralized local authorities ;
- public and parapublic sector companies".

---

[135]*Ibid.*

[136]Landry NGONO TSIMI, *op. cit,* p. 14.

This provision makes the Chambre des Comptes, together with its lower courts, the guarantor of financial and budgetary orthodoxy within communes and regions. However, it should be emphasized that, unless they have interfered in the handling of funds, authorizing officers are not concerned here. In fact, they are governed by the provisions of law no. 74/018 of December 05 1974 on the control of authorizing officers, managers and administrators of public funds, amended and supplemented by law no. 76/4 of July 08 1976.

The reason for this exclusion is that the legislator did not want the audit judge to slide into political control of the actions of local elected representatives, under the guise of a purely technical control of regularity. Under the terms of article 5 paragraph 2 of the aforementioned law no. 2003/005 of April 21, 2003, only the following are concerned:

- treasury accountants ;
- domain accountants ;
- municipal receivers, insofar as municipal revenues are managed by persons other than treasury accountants;
- subject accountants and all those designated as such.

The Chambre des Comptes is organized into sections, one of which is specifically responsible for "auditing and judging the accountants of decentralized local authorities, subject to the powers vested in the lower courts of audit".[137]

In conclusion, the construction and completion of the decentralization process in Cameroon involves several players, all of equal importance and whose roles are distributed in specific ways. At the top of the chain are the central and decentralized players, the State's supervisory authority over

[137] *Ibid.*

local players, who are at the heart of the implementation process, while the technical side is handled by specialized institutions responsible for monitoring, control and support. After this meticulous analysis of the institutional and legal framework of the decentralization process in Cameroon, it is now time to move on to our specific case study, in order to gauge the extent to which certain CTDs are in line with the dynamics illustrated above.

PART II: THE COMMUNE DE YAOUNDE D'ARRONDISSEMENT DE YAOUNDE II AT THE TEST OF DECENTRALIZATION: BETWEEN LOCAL DEVELOPMENT AND INTERNATIONAL COOPERATION

The impetus having been given by the central State, the CTDs in general, and the communes in particular, are now expected to be major development players, while at the same time fitting into the legal-institutional framework described above. The aim of our research is therefore to assess the extent to which Cameroon's communes are in line with the process of strengthening decentralization. This section will therefore focus on our study framework, the Commune d'Arrondissement de Yaoundé II. To this end, the first chapter gives a kind of general presentation or monograph of the structure we are studying (Chapter III), while the second evaluates the implementation of various local development and international cooperation policies (Chapter IV), the degree of effectiveness of which is a rating indicator of the structure's anchorage in the decentralization process.

CHAPTER III: PRESENTATION OF THE STUDY FRAMEWORK: MONOGRAPH ON THE ARRONDISSEMENT COMMUNE OF YAOUNDE II

Created by Presidential Decree *N°87/1365 of September 25, 1987*, the Commune de Yaoundé 2 was not operational until August 1988[138] . The current Commune d'Arrondissement of Yaoundé II has undergone several changes: first, when it was created, it was known as the Commune de Yaoundé II, according to the aforementioned decree, then as the Commune Urbaine de Yaoundé by decree *N°93/321 of November 25, 1993*, and today as the Commune d'Arrondissement de Yaoundé II after its break-up, which saw the birth alongside it of the Commune d'Arrondissement de Yaoundé VII[139] . Today, the Commune d'Arrondissement de Yaoundé 2 (CAY2) is one of seven Communes d'Arrondissement in the city of Yaoundé, the political capital of Cameroon. It covers an area of 22 km². Located between 45ème degrees north latitude and 15ème degrees south latitude, the Yaoundé 2 Commune d'Arrondissement, whose Town Hall is in the Tsinga district, is considered the gateway to and from Cameroon for all the world's leading figures, thanks to its proximity to the Presidential Palace (Palais de l'Unité). This gateway status is further enhanced by the presence of the sumptuous Palais des Congrès, the venue par excellence for national and international meetings held in Cameroon. This chapter begins with a presentation of the geographical and socio-economic aspects of the Commune d'Arrondissement de Yaoundé II (Section I), followed by an outline of the missions, operations and resources of the Commune d'Arrondissement de Yaoundé II (Section II).

---

[138] CAYII, *CAYII presentation* document, revised and updated version, December 2020, p. 1
[139] *Ibid.*

SECTION I: GEOGRAPHICAL AND SOCIO-ECONOMIC ASPECTS OF THE YAOUNDE II ARRONDISSEMENT COMMUNE

In this section, we will focus on the geographical (Paragraph I) and socio-economic (Paragraph II) aspects of the Commune d'Arrondissement of Yaoundé II.

PARAGRAPH I: GEOGRAPHICAL ASPECTS OF THE YAOUNDE II ARRONDISSEMENT MUNICIPALITY

In this section, we will focus on presenting the topography and climate of the Yaoundé II Arrondissement commune, as well as its geographical delimitation.

A - CLIMATE AND TOPOGRAPHY

The Commune d'Arrondissement of Yaoundé 2, some 270 km from the Atlantic Ocean, has a relief dominated by mountains, the most notable of which are[140] :

- Mont Mbankolo proudly displays the large John Paul II Auditorium, a place of prayer, meditation and spiritual awakening;
- Mont Fébé, with the Mont Fébé hotel, much appreciated for its golf course: the only one in Yaoundé, and the Benedictine monastery;
- Mont Messa, whose status as a green zone has just been reaffirmed by the government, is eager to exploit its many assets;
- NkolNyada, at the summit of which stands the majestic Palais des Congrès de Yaoundé. Yaoundé 2 is an urban Commune with a rural area covering around 15% of its surface area.

[140] CAYII, *op. cit*, p.2

The climate is equatorial, with two rainy seasons and two dry seasons whose alternation over time has been greatly disrupted, hence the term "Yaoundé-type" equatorial climate[141] .

B - GEOGRAPHICAL DELIMITATION

CAY2 is limited:

- to the north and northwest by the CAY1 ;
- to the south by the CAY6 ;
- to the south-west and south-east by CAY7; to the east by CAY3

PARAGRAPH II: SOCIO-ECONOMIC ASPECTS OF THE YAOUNDE II DISTRICT MUNICIPALITY

In this section, we will first present the demographic configuration of the Commune d'Arrondissement of Yaoundé II, then illustrate the local infrastructural and economic dynamics.

A - DEMOGRAPHICS OF THE COMMUNE

Demographically speaking, CAY2 is a cosmopolitan municipality characterized by peaceful cohabitation between its populations despite their diverse origins. Its population is estimated at just over 238,927, according to the 2005 general population census[142] , and is distributed across the 29 districts of its municipality according to the following table:

Table 1: List of different districts in the Commune d'Arrondissement of the Commune d'Arrondissement of Yaoundé II

[141] *Ibid.*

[142] CAY II, *Ibid*, p. 3

1	AZEGUE	16	ANGONO, DOUMASSI, PLATEAU
2	FEBE	17	CITE VERTE CAMP SIC
3	GRAND MESSA & ADMINISTRATIVE MESSA	18	EKOUDOU A
4	MADAGASCAR A (CAMP SIC)	19	BRIQUETERIE A
5	MADAGASCAR B (TONEAU)	20	BRIQUETERIE B
6	MESSA-MEZALA	21	EKOAZON, NKOABA'A
7	MOKOLO WALKS	22	TSINGA 2
8	MOKOLO DISTRICT A	23	MONT MESSA 3
9	MOKOLO DISTRICT B	24	TSINGA 1
10	NKOMKANA 1 & 3	25	MONT MESSA 1B
11	NKOMKANA 2	26	MONT MESSA 2
12	NTOUGOU 2 A	27	CITE VERTE SUD
13	NTOUGOU 2 B	28	MONT MESSA 1A
14	NTOUGOU I	29	EKOUDOU B
15	OLIGA		

(Source: CAYII presentation document, p 3-4)

B - LOCAL INFRASTRUCTURE AND ECONOMIC DYNAMICS

Yaoundé II is home to the Tsinga Mosque (the largest and busiest in Yaoundé), renowned Catholic Missions (Tsinga, Mokolo, Auditorium Jean Paul II), Protestant churches and other revivalist churches. On the health front, our commune is home to a number of health facilities, although it does not have one of its own. These include the Yaoundé Central Hospital, the largest in the country, and the Cité Verte District Hospital. However, the commune is in charge of building an integrated health center for one of its neighborhoods. In this case, the Briqueterie. In this respect, it is experiencing numerous difficulties in completing the work. Worse still, it is facing the real question of its equipment (furniture, complete hospital equipment). Economic activity in CAY2 is based on services, commerce, small trades and crafts. In this respect, it's worth pointing out that CAY2 is home to the Mokolo market, the largest in Yaoundé. Tsinga is also home to the crafts market. Despite all these good points, the informal sector still accounts for over 70% of economic activity[143] . Since 2010, the Yaoundé 2 Borough Council has been responsible for basic education, thanks to the application of the law on decentralization. In most cases, these powers are not accompanied by the means to implement them. As a result, the municipality has inherited a totally degraded school map, with schools in a state of disrepair and lacking everything. This situation is all the more worrying given that this type of education is free. As far as integrating young people into the production chain is concerned, the lack of vocational training is a real hindrance to both qualification and self-employment.

[143] CAYII, *Ibid* ,p. 4

SECTION II: MAIN MISSIONS, OPERATIONS AND RESOURCES OF THE YAOUNDE II MAYOR'S OFFICE

In this section, we first present the functioning of the Yaoundé II Commune d'Arrondissement (Paragraph I), and then analyze its development policy (Paragraph II).

PARAGRAPHI: OPERATION OF THE ARRONDISSEMENT COMMUNE OF YAOUNDE II

In this section, we will focus on the presentation of the municipal executive and other departments of the Commune, and on the Commune's various resources.

A - THE MUNICIPAL EXECUTIVE AND OTHER DEPARTMENTS

The deliberative body of CAY2 is the Municipal Council, made up of 41 councillors, including the municipal executive (the mayor and his 04 deputies) elected by the council and 05 grand councillors appointed by the mayor, who also sit on the Community Council (the deliberative body of the Communauté Urbaine de Yaoundé).

To carry out its duties, the Municipal Council has four commissions:

- Finance Committee;
- Projects and Cooperation Commission ;
- The Social Affairs Commission ;
- The Infrastructure and Major Works Commission.

The day-to-day running of CAY2 is ensured by the municipal executive led by the mayor, his 04 deputies and all the departments.

The organizational chart of the Town Hall, the headquarters of CAY2, includes the following main components:

- The municipal executive and its related departments (Mayor, Deputy Mayors, Communication Unit, Local Development and Decentralized Cooperation Support Unit, Mayor's Office, Accounting-Material);
- The General Secretariat, which includes the other departments;
- La Recette Municipale.

B - THE COMMUNE'S VARIOUS RESOURCES

Several types of resources were mobilized to implement and operate the DACs:

In addition to the commune's personal involvement through its own funds, the project has received strong financial, technical, material and human support from its partners: MINDUH (youth employment, various training courses and donation of a paving stone manufacturing unit), MINSANTE (vector and cholera control), CUY (transfer of 3,000m^2 of land for the construction of the Mairie and subsidy granted for the completion of work, other multi-faceted support), CREPA Cameroun and CAWST, an NGO based in Canada (potabilization of water at home, popularization of low-cost technologies), the NGOs ASSEJA, ASSOAL (awareness-raising, training, neighborhood development projects, financing and support for project implementation), Colombes Town Hall in France (construction of an autonomous mini gravity-fed drinking water supply network with 13 community standpipes scattered throughout the

Messa-Carrière neighborhood, installation of a sanitation network, and cultural and sporting exchanges)[144] ;

PARAGRAPH II: DEVELOPMENT POLICY FOR THE ARRONDISSEMENT OF YAOUNDE II

In this section, we'll be highlighting the various activities carried out by the municipality, as well as the ways in which the main thrusts of its policy are implemented.

A - MAIN ACTIVITIES OF THE ARRONDISSEMENT OF YAOUNDE II

The CAY2, whose main mission is to ensure the well-being of its populations, is therefore the driving force behind grassroots development. To this end, in addition to its regular tasks of issuing civil status certificates, CAY2, through its executive, has resolutely embarked on a number of projects whose importance and impact are already beginning to set it apart from its sister communes.

The main lines of action for the Commune of Yaoundé II are[145] :

- Hygiene and cleanliness (Wednesday has been declared a cleanliness day throughout the Commune);
- Vector control ;
- Developing self-employment;
- Youth employment, where the High Labor Intensity program has already made its mark through visible and appreciable achievements;

[144] CAY II, *Ibid*, p. 7

[145] CAYII, *Ibid*, p. 6

- Information, awareness-raising and training;
- Development of basic infrastructure ;
- Social engineering;
- Supporting people ;
- The participatory budget ;
- Decentralized cooperation and local development.

B - THE POLICY FOR IMPLEMENTING THE MAIN ROUTES

Given the limited resources available, certain policies have been developed by the CAY executive[146] :

- The creation of a Local Development and Decentralized Cooperation Support Unit within the Commune, headed by an engineer experienced in development and cooperation issues. This unit is the focal point for the development of the municipal executive's policy, and serves as an interlocutor with all the Commune's potential partners;
- The creation of Animation and Development Committees (CAD), with one per district. The CADs bring the Commune closer to its residents. They are the Commune's relays in the neighborhoods. They are structured around a board appointed by the local people themselves, and councillors including the Chef de quartiers, chefs de blocs and municipal councillors living in each district.

Under the coordination of the mayor's office, the CAD's mission is to lead development projects in the neighborhoods, supervising the

[146] CAY II, *Ibid*. p. 6

masses, informing the population and the mayor's office, and developing partnerships within the neighborhoods with NGOs and other partner organizations.

- The development of internal cooperation. In this respect, thanks to financial support from the Fond d'Aide Intercommunale (FEICOM), we have begun work on the construction of our town hall, which will open in June 2012. In addition, thanks to our partnerships with local NGOs (ASSEJA, ASSOAL, etc.), we have so far carried out just over 25 projects and conducted more than 30 seminars and training workshops for the benefit of our populations. We have also benefited from the Ministry of Public Health and the Ministry of Urban Development with a project to combat vectors within the framework of sanitation for the former, and youth training and self-employment facilities for our populations for the latter.
- Development of external cooperation. The Centre Régional pour l'Eau Potable et l'Assainissement à faible coût (CREPA), the Canadian NGO CAWST, the Mairie de Colombes in France and the SIAAP in France are perfect illustrations of this, although we still expect a great deal from these partnerships.

All in all, the Commune d'Arrondissement de Yaoundé II, in view of the dynamism of its executive, ranks as one of the most dynamic local authorities in the City of Yaoundé. Due to its geographical position, its socio-economic and infrastructural configuration, it is the most likely to considerably boost the development of the city of Yaoundé. The participatory development policy and the diversification of partners have led to an increase in the commune's resources. Nevertheless, beyond a dynamic municipal executive, a favourable spatial framework and relatively available resources, it would now be appropriate to question the effective implementation of local development and decentralized cooperation within the Yaoundé II Arrondissement Commune, in order to assess the extent to which it is in line with the central government's drive to strengthen decentralization.

CHAPTER IV: THE ARRONDISSEMENT MUNICIPALITY OF YAOUNDE II: THE TEST OF LOCAL DEVELOPMENT AND INTERNATIONAL COOPERATION

The implementation of decentralization in Cameroon offers new opportunities for local authorities to promote their socio-economic development without waiting for the "manna" from the central state[147] . For its part, CAY 2 is now the driving force behind grassroots development in the Yaoundé 2 district. To this end, several policies have been developed under the impetus of the municipal executive. These are characterized by the formulation of development objectives (Section I) and the creation and implementation of structures to stimulate the local economy (Section II).

SECTION I: FORMULATING AND IMPLEMENTING LOCAL DEVELOPMENT AND INTERNATIONAL COOPERATION OBJECTIVES

In this section, we will test the effectiveness of local development (Paragraph I) and then evaluate the relevance of decentralized cooperation (Paragraph II) within the Yaoundé II Commune d'arrondissement.

PARAGRAPH I: THE OPERATIONALIZATION OF LOCAL DEVELOPMENT AS AN EFFECTIVE POLICY IN CAYII

In this paragraph, in order to assess the level of operationalization of local development within CAY II, we will analyze the two main structures created for this purpose, namely the local development and decentralized cooperation support unit (CADLCD) and the animation and development committees (CAD).

[147]Muriel SAME EKOBO and Olivier IYEBI MANDJEK, *op. cit*, p. 9

A - THE CREATION OF A LOCAL DEVELOPMENT AND DECENTRALIZED COOPERATION SUPPORT UNIT (CADLCD) WITHIN THE COMMUNE.

This unit is the focal point of the municipal executive's development policy. It supports the commune in the decentralization process, in setting up projects, and in the search for funding[148] . It also acts as an interface with all the commune's potential partners, and is responsible for developing the commune's decentralized cooperation. CADLCD is also the mayor's focal point for the local population. It is headed by an engineer with extensive experience of development and cooperation issues. Alongside this, we also have a communications unit that informs and animates the masses in order to keep them alert to essential issues; a technical department for urban planning and development and a hygiene department that also works on a daily basis to meet the development challenges of the Yaoundé 2 locality.

B - THE CREATION OF ANIMATION AND DEVELOPMENT COMMITTEES (CADS) AS A MODEL FOR LOCAL GOVERNANCE.

With a view to supporting and involving the people of Yaoundé 2 in improving their living conditions through self-help solutions, the current municipal executive has set up the "Comité d'Animation et de Développement" (Animation and Development Committees)[149] , veritable development hubs in the neighborhoods of Yaoundé II. The aim here was to get the population involved and participate in the commune's development initiatives. For Yaoundé 2, the challenge has always been to

---

[148] CAYII, *op. cit*, p. 8
[149] *Ibid.*

implement a participatory approach in all actions undertaken by the commune. To this end, after a series of field visits to all Yaoundé 2 neighbourhoods in 2007 by the municipal executive, accompanied by the various municipal departments and partner NGOs, to make contact and talk to the population about their vision of the commune, the mayor received a number of complaints:

- No local development projects in place;
- Absence of the population in the decision-making processes and in the projects or actions of the town hall;
- Non-results-oriented management ;
- City Hall services considered to be the executioners of the population in all areas;
- Lack of proximity between the town hall and the population ;
- Lack of awareness of the Commune's role ;
- Ignorance of the municipal executive.

In December 2007, a resolution authorizing the creation of the CADs was presented to the council by the Mayor and voted unanimously by the councillors[150] ; In April 2008, the then Prefect of Mfoundi, Mr. BETI ASSOMO, proceeded to install the CADs with the distribution of working equipment and a symbolic operating fund[151] ; October 2010 saw the effective implementation of the 18 Yaoundé II neighborhood development plans[152] ; Similarly, the Participatory Budgeting process has been underway since October 2008 in the commune[153] . To date, the CADs have

[150] CAYII, *Ibid*, p. 10
[151]*Ibid.*
[152]*Ibid.*
[153]*Ibid.*

formed a network to ensure that their voice is heard. They will now be able to take part in the municipal council.

PARAGRAPH II: DEVELOPING COOPERATION WITH INTERNATIONAL DEVELOPMENT PARTNERS

In this section, which aims to assess the effectiveness of international cooperation within the Commune d'Arrondissement de Yaoundé II, we will first review the Commune's various international partners, and then highlight the various actions carried out as part of this international cooperation.

A - INTERNATIONAL PARTNERS OF THE ARRONDISSEMENT OF YAOUNDE II

In terms of external cooperation, CAY 2 has established excellent relations with the NGO Eau et Assainissement pour l'Afrique (EAA), the Canadian NGO CAWST, the Colombes Town Hall in France, the SIAAP in France, the Franckental Municipality in Germany, and the BUMBU Municipality in Kinshasa[154] . Numerous exchanges have also taken place with World Bank structures.

B - INTERNATIONAL COOPERATION INITIATIVES IN CAYII

Turning to decentralized cooperation in the strict sense of the term, the most significant cooperation to date has been between the Yaoundé 2 commune in Cameroon, the Colombes town council in France and the SIAAP (Syndicat Interdépartemental pour l'Assainissement de l'Agglomération Parisienne). The cooperation between these three parties has seen the implementation of a drinking water supply project for the

---

[154] CAYII, *Ibid.* 12

people of Messa-Carrière, through the installation of a water catchment on the flank of Mount Messa, a natural and organic water treatment system, storage in low-cost structures (03 ferrocement tanks of 25 m3 each) and, above all, free, permanent gravity-fed distribution of this precious sesame, which was so lacking in the population of this district[155] .

The continuation of this decentralized cooperation has given rise to a second project, this time to increase water storage capacity, extend the network and, above all, improve sanitation[156] . Work has already begun on this project. Emphasis is placed on social engineering, with the noble aim of ensuring that beneficiaries take ownership of the project. The local management committee for this important project has instituted a contribution of *2,400 CFA francs per household per year* as a fund for equipment maintenance and, above all, as a personal contribution from this district to serve as leverage for contributions to future projects[157] .

On a completely different level, the commune of Yaoundé 2 maintains exchanges and the sharing of experience in local governance with the Bourg de Bumbu, one of the communes of the city of Kinshasa in the DRC[158] . The emphasis here is on the transferability of a successful initiative to other communes in need of better ways of involving their populations in the management of the common good. After all these considerations, it is also appropriate to mention the limits and some prospects relating to the local development and international cooperation policy of the Yaoundé II Commune d'Arrondissement.

---

[155]*Ibid.*
[156]*Ibid.*
[157]CAYII,*Ibid,* p.16.
[158] *Ibid.*

SECTION II: LIMITS AND PROSPECTS OF LOCAL DEVELOPMENT AND INTERNATIONAL COOPERATION POLICIES IN THE ARRONDISSEMENT OF YAOUNDE II

This section will focus on the limits and prospects of the local development and international cooperation policies of the Yaoundé II Commune d'Arrondissement. To do so, we will first highlight the difficulties associated with their implementation (Paragraph I), and then see to what extent the consolidation of achievements is the main means of improving prospects (Paragraph II).

PARAGRAPH I: MAIN DIFFICULTIES INHERENT IN THE WORK OF THE ARRONDISSEMENT MUNICIPALITY OF YAOUNDE II IN TERMS OF LOCAL DEVELOPMENT AND INTERNATIONAL COOPERATION

With regard to the difficulties associated with CAY II's action in terms of local development and international cooperation, a distinction is made between structural difficulties linked to CAYII and those external to CAYII.

A - STRUCTURAL LIMITS TO CAYII

The difficulties are manifold. Without being exhaustive, we can mention[159] :

- Insufficient local staff to deal with development issues, as well as the capacity of existing staff;

[159]*Interview with the Head of Technical Services at CAY II on June 20, 2020*

- The lack of equipment at the integrated health center is a serious obstacle to setting up a communal health insurance scheme for vulnerable people;
- The growing need for basic education;
- The lack of qualifications among DAC managers;
- Low local revenues;
- The lack of logistical and material resources allocated to DACs ;
- The Commune's Local Development and Decentralized Cooperation Support Unit (Cellule d'Appui au Développement Local et à la Coopération Décentralisée) is poorly equipped;

B - EXTERNAL LIMITS OF THE CAYII

The main external limitations of the Yaoundé II Commune d'Arrondissement's action in terms of local development and international cooperation are the lack of institutional support from central government, and the timid transfer of powers, which is nevertheless taking place gradually[160] . Despite these difficulties, it is nevertheless possible to foresee a brighter future for the CAYII, provided certain conditions are met and certain achievements are consolidated

[160]Landry MEPUI ABAH, *op. cit*, p.11
[160] Yannick Félix PEGUI, *op. cit*, p. 30

PARAGRAPH II: THE NEED TO PRESERVE WHAT HAS ALREADY BEEN ACHIEVED AND TO REINFORCE THE ACTIONS UNDERTAKEN

If the Yaounde II Commune is to strengthen its actions in the fields of local development and international cooperation, it will need to preserve certain achievements. First and foremost, the need to build up a genuine local development engineering system, and to optimize previous initiatives.

A- THE NEED FOR GENUINE LOCAL DEVELOPMENT ENGINEERING

Faced with the challenges posed today by numerous external heteronomous forces (including globalization, the territorial attractiveness of companies and people...), territories are more than just a mute spatial framework, they are now focused on promoting and popularizing their socio-economic fabric. As a result, they must constantly work to renew their competitive advantages[161] . In addition, they need to increase their potential for creating resources and organizational skills: "*It's a commitment to constant innovation, to anticipating rather than passively adapting*"[162] . In Cameroon in particular, the introduction of decentralization is opening up new opportunities for local authorities (communes, regions, cities, etc.) to promote their economic, social and cultural development. For its part, the district of Yaoundé 2, with its geographical position, density and economic organization, should "follow in the *footsteps" of* this dynamic of aptitude, which in the "*age of global cities"* is imposed on all territories[163] . This appears to be a challenge to

[161] *Ibid.*
[162] *Ibid.*
[163] Yannick Félix PEGUI, *Ibid,* p. 30.

the commune to implement a genuine strategy of territorialized development. This strategy revolves around a number of axes, including the promotion of business incubators and the construction of a genuine local development engineering system. In particular, the creation and implementation of structures and tools for local development, over and above those already existing in the commune. By way of example, we suggest the creation of a local statistics and economic studies observatory (OLSEE) within the town hall. This structure could make a useful contribution to understanding the local economy and act as a strategic territorial watchdog. In this way, it will be able to provide all the necessary information prior to the actions of the municipal executive, particularly in terms of promoting the local economy.

B - OTHER RECOMMENDATIONS FOR OPTIMIZING THE WORK OF THE YAOUNDE II DISTRICT COUNCIL

As part of a strategic approach to development, the Town Hall needs to put in place mechanisms to increase the participation of local stakeholders in the search for consensual solutions to local development problems (through forums, training seminars, participatory consultations with local populations on projects, etc.). Publicizing CAY 2's experiences via the Internet portal and the local press.

In addition, improving the supply of basic infrastructure and services is a necessity in order to boost the economic attractiveness of the area, open up working-class neighborhoods and improve the social environment.

The need to raise the skill level of local staff and recruit personnel specialized in specific areas of activity, so that the commune can take full advantage of the potential offered by decentralization. For it is said that the establishment of a strong and competent local government is an essential condition for the smooth running of the commune. Finally, the importance of setting up structures to support the creation and sustainability of businesses (support for micro-projects, policies to encourage young entrepreneurship, etc.). As well as the importance of multiplying

Commune both internally and externally, in order to benefit from the experience of certain players and mobilize more resources.

GENERAL CONCLUSION

Ultimately, our work was entitled: ***"Local development policies and international cooperation within Cameroonian communes in a context of strengthening decentralization: the case of the Yaoundé II Commune d'Arrondissement",*** and its aim was to analyze the extent to which Cameroonian communes are tied in with the ongoing decentralization process. For this purpose, we used the Yaoundé II Commune d'Arrondissement as a case study. After a conceptual clarification that enabled us to precisely circumscribe the key notions of our topic, we were able to identify the main question of our subject, which was **In the current context of strengthening decentralization, can we speak of the implementation of local development policies and international cooperation in the Commune d'Arrondissement of Yaounde II?** Answering this question led us to break down our argument into four main chapters with related themes. The first chapter reviews the evolution of the legal framework for decentralization in Cameroon, while the second focuses on the different actors involved in this process. While the third chapter provided a monograph of the study framework, the fourth assessed the effectiveness of local development and international cooperation policies within the Yaoundé II Commune d'Arrondissement, while examining the different perspectives. At the end of our investigation, materialized by a six-month internship within our study framework, we were thus able to come into contact with the current realities of local administration. Overall, we found that, although Cameroon has adopted decentralization as its official mode of territorial administration since the adoption of the Constitution of January 18, 1996, it was not until the enactment of the 2004 laws that the legal and institutional framework really began to take shape, and was completed with the enactment of the General Code of the CTD and the establishment of the Regional Councils. As for the degree to which the CTDs are in tune with this dynamic, our foray into the Commune d'Arrondissement de Yaoundé II shows us that, while the will is there, the limited means often available, and the timidity of the transfer of powers, can limit the action of the CTDs. Nonetheless, to strengthen the action of the CTDs, we need to preserve certain achievements and reinforce the support of the central government.

BIBLIOGRAPHY

A. General works

BACHELARD Gaston, *Le nouvel esprit scientifique* (Paris: Les Presses universitaires de Paris, France).

France, 10th edition, 1968, p. 131.

BEAUD Michel, *L'art de la thèse*, Paris, La Découverte, revised edition, 2006 p. 27.

BECKER Howard, *Les ficelles du métier*, Paris, Guides Repères, La Découverte, 2002,
p.180.

BOURDIEU Pierre, *Propos sur le champ politique*, Presse Universitaire de Lyon, Lyon, 2000, p. 17

GRAWITZ Madeleine, *Méthodes des sciences sociales,* Paris, 11th edition, Dalloz, 2001,
p.53.

OLIVIER Laurence, BEDARD Guy and FERRON Julie, *L'élaboration d'une problématique de recherche : sources, outils et méthodes*. L'Harmattan, Logique sociale, 2005, p. 28.

QUIVY Raymond, Van CAMPENHOUDT Luc, *Manuel de recherche en sciences sociales*, Paris, Dunod, 2nd edn, 1995, pp.98-100.

B. Specialized books

BATTISTELLADario, PETITEVILLE Franck and VENNESSON Pascal, *Dictionnaire des relations internationales*, 3rd edition, Dalloz

BERGER P., LUCKMANN Th., *The Social Construction of Reality*, New York, Anchor, 1966

CRANE D. (ed.), *The Sociology of Culture*, Oxford, Blackwell, 1994, pp. 117-153.

CRONIN Bruce, *Community under anarchy. Transnational identity and the evolution of cooperation*, New York, Columbia University Press, 1999.

CHARILLON Frédéric (dir.), *Politique étrangère nouveaux regards*, Paris, Presses de France.
Sciences po, 2006, p.13

DEVIN Guillaume, *Sociologie des relations internationales*, Collections Repères
n°335, march 2018, 128p

DEUTSCH Karl,*SeminaronPolitical community and the North Atlantic: international organization in the light of historical experience*,1957.

DIAMOND Louise and MACDONALD John, *Multi-track Diplomacy, A system
approach to peace*, Kumarian Press, 3rd edition, 1993, 182p.

DOBBIN F., *Forging Industrial Policy*, Cambridge, Cambridge University Press,
1994.

FLIGSTEIN N., *The Transformation of Corporate Control*, Cambridge, Harvard University Press, 1990

GRAWITZ Madeleine, LECA Jean and THOENIG Jean Claude, *Les Politiques publiques*, Tome 4, Paris, PUF, 1985

GRAY John, *False Dawn: The Delusions of Global Capitalism*, London and New York, Granta Books and New Press, 1998, p 120.

JONES Charles O., *An introduction to the study of public policy*, Belmont (California), Duxbury Press, 1970

KAUL, INGE, GRUNBERG Isabelle and STERN Marc, *Les biens publics mondiaux : La coopération internationale au XXIe siècle,* Paris, Economica, 2002

KLÜBER Daniel et de MAILLARD Jacques ,*Analyser les politiques publiques*, Collection Politique en plus, Presses universitaires de Grenoble, p. 8.

LASSWELL Harold, *Who gets what, when and how* (Cleveland, Ohio: Meridian Books, 1936).

LATOUCHE Serges (ed.), *L'économie dévoilée. Du budget familial aux contraintes planétaires*, Paris, Les Éditions Autrement, coll. Mutations, n° 159, 1995, p. 190-195.

LOWI Theodore, *The End of Liberalism*, New York, N. Y., Norton, 1969.

LUTZ George and LINDER Wolf, *Structures traditionnelles dans la gouvernance locale pour le développement local,* University of Berne, Institute of Political Science, Switzerland, 2004.

MENY Yves and THOENIG Jean-Claude, *Politiques publiques*, Paris, PUF, 1989

MEPUI ABAH Landry, *Dynamiques des politiques décentralisées au Cameroun, Analyse des enjeux et défis sociopolitiques économiques et culturels, L'Harmattan,* Paris, 2012, 115p.

MEYER J.W., SCOTT W.R., *Organizational Environments. Ritual and Rationality*, Beverly Hills, Sage, 1983.

MULLER Pierre, *Les Politiques publiques*, Paris, PUF, Coll. "Que sais-je", 2009, 8th ed.

PAQUIN Stéphane, BERNIER Luc and LACHAPELLE Guy (dirs.), *L'analyse des politiques publiques*, Les Presses de l'Université de Montréal, 2010, p. 8.

PETERS Guy, *American Public Policy*, Basingstoke, Mac Millan, 1986

POWELL.W, DIMAGGIO P. (eds), *The New Institutionalism in Organizational Analysis*, Chicago, University of Chicago Press, 1991, pp. 1-40.

SACHS W. and GUSTAVO E., *Des ruines du développement*, Montréal, Écosociété, 1996, 138 p

SCOTT Jervis, *Realism, Neoliberalism and cooperation understanding the debate in International Security,* 24(1), 1999, p. 42 - 63

SCOTT W. R., MEYER J.W. et al, *Institutional Environments and Organizations*, Thousand Oaks, Sage, 1994, chapters 11 and 12.

SOYSAL Y., *Limits of Citizenship*, Chicago, University of Chicago Press, 1994.

C. Articles

AMOUGOU Thierry, Le Nouveau Paradigme de la Coopération au Développement (NPCD), Quels enjeux pour le développement des pays partenaires, in *Economie et Solidarités*, Volume 40(n°1-2), p. 63-83.

GONTCHAROFF George, Le développement local : petite généalogie historique et conceptuelle, in *Territoires*, Octobres 2002, p. 1-3.

HECLO Hugh, "Issue Network and The Executive Establishment" in Anthong King (ed.), *The New American Political System*, Washington DC, American Enterprise Institute, 1978

HOCKING Brian, Catalytic Diplomacy: Beyond 'Newness' and 'Decline', in *Innovation in Diplomatic Practice*, p.21-42

HUSSON Bernard, le développement local, in *Agridoc n°1*, available on : www.resacoop.org

LAKE David A., "British and American Hegemony Compared", *in* Jeffrey A. Frieden and David A. Lake, *International Political Economy. Perspectives on Global Power andWealth*, Londonand New York, Routledge, 1997, p. 129.

MEARSHEIMER John J., "The False Promise of International Institutions", *in International Security* 19-3 Winter 1994-1995, pp. 5-49.

MEBENGA Mathieu, "La fiscalité dans la gouvernance décentralisée au Cameroun", in *Enjeux*, n° 45-46, July 2012, p. 17

PEKASSA NDAM Gérard, "La place de l'administration dans les politiques de décentralisation et de la gouvernance locale au Cameroun", in *Enjeux*, n° 45-46, July 2012, p. 14

RUGGIE John Gerard, "Multilateralism: The Anatomy of an Institution", *in* John G. Ruggie, editor, *Multilateralism Matters: The Theory and Praxis of an InternationalForm*, New-York, Columbia University Press, p. 3.

SAME EKOBO Muriel and IYEBI MANDJEK Olivier, "Gouvernance territoriale et action publique ai Cameroun", in *Enjeux*, n°45-46, July 2012, p. 9

SCOTT Jervis, Realism, game theories and cooperation, in *Worlds Politics*, XL (3), 1988, pp. 317-349

STEIN Janice, "l'analyse de la politique étrangère: à la recherche de groupes de variables dépendantes et indépendantes", in *Etudes Internationales*, Vol II, N°3, 1971, p. 373;

TOUNA MAMA Ernest, "Qu'est ce que la mondialisation", in *La mondialisation et l'économie camerounaise,* Yaoundé, Saagraph and Friedrich Ebert Foundation, 1998.

D. Scientific work

- **Theses**

NGONO TSIMI Landry, *L'autonomie administrative et financière des collectivités territoriales et décentralisées : l'exemple du Cameroun*, Doctoral thesis defended at the Université de Paris-Est, 2010, p. 32

➢ **Memories**

BETANCUR RAMIREZ Santiago, *Quel rôle pour les gouvernements locaux sur la scène internationale? L'action internationale des collectivités locales entre la France et l'Amérique latine,* dissertation defended publicly for the Master II in Public Affairs, ENA, Paris, 2018, 171p.

PEGUI Yannick Félix, *Décentralisation et fonctionnement des Communes au Cameroun. Cas de la Commune d'Arrondissement de Yaounde II*, dissertation submitted publicly for the Master II degree in economics, University of Yaoundé II, Yaoundé, 2012

E. Official reports and documents

Commune d'Arrondissement de Yaoundé II, *official presentation document*, December 2020

Republic of Cameroon, *Constitution of January 18, 1996.*

F. Courses and lessons

NGONO TSIMI Landry- Cours de *Collectivités locales, décentralisation et droit de la coopérationdécentralisée au Cameroun*, Master "Coopération internationale, Action humanitaire et Développementdurable", Yaoundé, Institut des Relations internationales du Cameroun: 2015,

G. Websites

AUTON Yves, "Etudes internet et développement local, première partie : le développement local", 2000 , available at: www.admiroutes.asso.fr.

HUSSON Bernard, "La coopération Décentralisée : légitimer un espace publique local au Sud et à l'Est", 2000, available at www.capcoopération.org, last consultation date June 13, 2020.

www.pndp.org visited on January 24, 2021 between 5:00 pm and 6:00 pm.

TABLE OF CONTENTS

Printed by Books on Demand GmbH, Norderstedt / Germany